Findings from Production Management Research

Series Editor

Peter Burggräf, International Production Engineering & Management, Universität Siegen, Kreuztal, Nordrhein-Westfalen, Germany

In the dynamic realm of industrial operations, production management remains the core for organizational efficiency. This series is a collection of doctoral thesis delving into the latest insights and innovations shaping production management, offering a comprehensive exploration of emerging trends and best practices. From the integration of advanced digital technologies to the effects of sustainability initiatives, this series covers key facets driving the evolution of modern production management. Drawing upon cutting-edge research and practical case studies, this collection is an indispensable resource for industry professionals, academics and students seeking to navigate and capitalize on the dynamic landscape of modern production management.

Fabian Steinberg

Machine Learning-based Prediction of Missing Parts for Assembly

 Springer Vieweg

Fabian Steinberg
Universität Siegen
Siegen, Germany

Dissertation zur Erlangung des Grades eines Doktors der Ingenieurswissenschaften vorgelegt von Fabian Steinberg, M. Sc. eingereicht bei der Naturwissenschaftlich-Technischen Fakultät der Universität Siegen, Siegen 2024
Betreuer und erster Gutachter: Univ.-Prof. Dr.-Ing. Peter Burggräf, Universität Siegen
Zweiter Gutachter: Univ.-Prof. Dr. rer. nat. Jochen Garcke, Universität Bonn
Tag der mündlichen Prüfung: 09. Februar 2024

ISSN 3005-1649 ISSN 3005-1657 (electronic)
Findings from Production Management Research
ISBN 978-3-658-45032-8 ISBN 978-3-658-45033-5 (eBook)
https://doi.org/10.1007/978-3-658-45033-5

This Springer Vieweg imprint is published by the registered company Springer Fachmedien Wiesbaden GmbH, part of Springer Nature.
The registered company address is: Abraham-Lincoln-Str. 46, 65189 Wiesbaden, Germany

Acknowledgements

This thesis on the prediction of missing parts for assembly was written during my time as a research assistant at the Chair of International Production Engineering and Management (IPEM) at the University of Siegen, Germany. I would like to thank the people who contributed to the success of this work.

Special thanks go to Prof. Dr.-Ing. Peter Burggräf, holder of the IPEM chair, for the opportunity to do my doctorate and for his support, encouragement and trust in me as a person. The environment he created, the first-class education, the wealth of ideas and the collegial management style were ideal conditions for this work and for my personal development.

I would also like to thank Univ.-Prof. Dr. rer. nat. Jochen Garcke, Research Group Leader at the Institute for Numerical Simulation at the University of Bonn and Head of the Business Area Numerical Data-Driven Prediction at the Fraunhofer Institute for Algorithms and Scientific Computing SCAI, for his willingness to serve as the second expert witness. Dr.-Ing. Robert Brandt, holder of the Chair of Materials Systems for Lightweight Vehicle Construction at the University of Siegen, for assuming the chair of the doctoral committee, and Univ.-Prof. Dr.-Ing. Axel van Heel, holder of the Chair of Materials Science and Materials Testing, for being a member of the doctoral committee.

Many of my former and active colleagues at IPEM have had a decisive influence on the content of this work, and I am deeply grateful to them. The mutual helpfulness, the high motivation, the creativity and the friendship have always given me new motivation. I would like to emphasize the cooperation with Dr.-Ing. Johannes Wagner, Benjamin Heinbach, Till Saßmannshausen, Rene Sauer, Alexander Becher, Philipp Nettesheim and Maximilian Schütz. Furthermore, I would like to thank my student assistants and the students who wrote their bachelor and master theses under my supervision, and continuously supported me in my work at IPEM. I would also like to thank Mrs. Mailin Klaas for the first-class administrative organization and for her sympathetic support throughout my time at the chair.

My deepest and most heartfelt thanks go to my wife Kristina Bieker-Steinberg for her loving support, her motivating encouragement and the space she has given me, as well as to my parents Birgit and Andreas Steinberg for their unconditional support on my life's journey and for encouraging me to change my career from industry back to academia. Without their loving support, this thesis would not have been possible.

Olpe   Fabian Steinberg
February 2024

Abstract

Manufacturing companies are faced with the challenge of managing increasing process complexity, while at the same time having to meet ever higher demands in terms of on-time delivery and product costs. Especially at points in the value chain such as assembly, where different material flows converge, it is often not possible to provide the components required for an order in a timely and synchronized manner. Early identification of missing parts at the beginning of assembly can help to take countermeasures to meet the required delivery dates. To achieve this, this thesis develops machine learning based prediction models that can predict potential missing parts at the start of assembly at an early stage in the value chain. The development of the models was carried out as case studies at manufacturing companies in the machine industry. As a basis for the development, an extensive systematic literature search was conducted on existing approaches for the prediction of lead times of production orders. The result was that no approach exists that takes into account the full complexity of manufacturing companies. In particular, with regard to the data used, it became clear that information about the product to be manufactured—so-called material data—has not been used up to now. Based on the systemic review, a model for predicting missing parts from in-house production was implemented. It was shown that classification approaches achieve the best possible model quality for components from in-house production. With the defined modeling approach—classification—it was then verified that material data has a significant influence on the model quality and is therefore relevant for the prediction of missing parts at the start of assembly. Finally, a model for predicting delivery delays in the purchasing process was implemented, which makes it possible to predict potential missing parts from suppliers at the time of ordering. The case studies show that the use of machine learning for the prediction of missing parts in both in-house production and the purchasing process can identify delays in the start of assembly at an early stage. The developed models are therefore suitable as a support system for production planners and controllers as well as purchasing departments to improve material availability at the start of assembly.

Zusammenfassung

Produzierende Unternehmen stehen vor der Herausforderung, eine stetig wachsende Prozesskomplexität bei gleichzeitig steigenden Anforderungen an Termintreue und Produktkosten zu beherrschen. Insbesondere an Stellen der Wertschöpfungskette wie der Montage, an denen verschiedene Materialflüsse zusammenlaufen, gelingt es häufig nicht, die für einen Auftrag benötigten Komponenten rechtzeitig und synchron bereitzustellen. Das frühzeitige Erkennen von Verzögerungen bis zum Montagebeginn kann helfen, Gegenmaßnahmen einzuleiten, um die geforderten Liefertermine einzuhalten. Um dies zu erreichen, wurden im Rahmen dieser Arbeit maschinelle lernbasierte Prognosemodelle entwickelt, die potentielle Fehlteile zum Montagestart frühzeitig vorhersagen können. Die Entwicklung der Modelle wurde jeweils im Rahmen von Fallstudien bei produzierenden Unternehmen aus dem Maschinen- und Anlagenbau betrachtet. Als Grundlage für die Entwicklung wurde zunächst eine umfassende systematische Literaturrecherche zu bestehenden Ansätzen zur Vorhersage der Durchlaufzeit von Fertigungsaufträgen durchgeführt. Das Ergebnis war, dass bislang kein Ansatz existiert, der die gesamte Komplexität produzierender Unternehmen berücksichtigt. Insbesondere bei den verwendeten Daten zeigte sich, dass Informationen über das zu fertigende Produkt—so genannte Materialdaten—bisher nicht genutzt werden. Auf Basis dieser Untersuchungen wurde zunächst ein Modell zur Vorhersage von Fehlteilen aus der eigenen Produktion implementiert. Dabei zeigte sich, dass in diesem Bereich Klassifikationsansätze die bestmögliche Modellgüte erreichen. Mit der gewählten Modellierungsart—Klassifikation—wurde anschließend ermittelt, dass Materialdaten einen signifikanten Einfluss auf die Modellgüte haben und somit für die Vorhersage von Fehlteilen am Montagestart relevant sind. Schließlich wurde ein Modell zur Vorhersage von Lieferterminverzögerungen im Einkaufsprozess implementiert, mit dessen Hilfe potentielle Fehlteile von Lieferanten bereits zum Zeitpunkt der Bestellung vorhergesagt werden können. Die betrachteten Fallbeispiele zeigen, dass durch den Einsatz von maschinellem Lernen zur Vorhersage von Fehlteilen in der eigenen Fertigung sowie im Einkaufsprozess Ursachen für Verzögerungen des Montagestarts frühzeitig identifiziert werden können. Die entwickelten Modelle eignen sich somit als Assistenzsystem für Produktionsplaner und -steuerer sowie Einkaufsabteilungen, um die Materialverfügbarkeit zum Montagestart zu verbessern.

Contents

List of Abbreviations

€	Euro
AB	Adaptive Boosting
AdaBoost	Adaptive Boosting
ANN	Artificial Neural Networks
APS	Advanced Planning System
ASD	Assembly Start Delayer
BAG-DT	Bagged Trees
CAD	Computer Aided Drawing
CO	Combinatorial Optimization
CON	Constant
CRISP-DM	Cross Industry Standard Process for Data Mining
CSV	Comma Separated Value
CT	Control Theory
DecT	Decision Theory
DT	Decision Tree
ERP	Enterprise Resource Planning
GB	Gradient Boosting
H	Heuristics
JIQ	Jobs in Queue
KNN	K-Nearest-Neighbor
LINREG	Linear Regression
LOGREG	Logistic Regression
LP	Linear Programming
LR	Linear Regression

LSA	Latent Semantic Analysis
MES	Manufacturing Execution Systems
ML	Machine Learning
MLP	Multilayer Perceptron
NC	Numerical Control
NLP	Nonlinear Programming
NOP	Number of Operations
OP	Operation
OR	Operations Research
PC	Principal Component
PCA	Principal Component Analysis
PDM	Precedence Diagram Method
PLM	Product Lifecycle Management
PPC	Production Planning and Control
QC	Quality Criteria
QT	Queuing Theory
RAN	Random
ReLU	Rectified linear unit
RF	Random Forest
RQ	Research Question
SCM	Supply Chain Management
SVC	Support Vector Classifier
SVM	Support Vector Machine
tanh	Hyperbolic Tangent
TU	Time Units
TWK	Total Work
WIQ	Work in Queue

List of Symbols

ASD	Assembly Start Delayer
CD	Completion Date
$DD_{finished}$	Finished Date Deviation
DDL	Delivery Date Lateness
DD_{rel}	Relative Date Deviation
DD_{start}	Start Date Deviation
DOF_{actual}	Actual Finished Date of an Order or Operation
DOF_{target}	Target Finished Date of an Order or Operation
DOS_{actual}	Actual Start Date of an Order or Operation
DOS_{target}	Target Start Date of an Order or Operation
F	F-score
FN	False Positive
FP	False Negative
$G_{left/right}$	Impurity of the Left/Right Subset
I1	Inital Inventory
I2	Final Inventory
IC	Completed Inventory
ID	Disrupted Inventory
IP	Inventory in Process
IW	Waiting Inventory
J(k,t)	Cost Function
LT	Lead Time
LT_{actual}	Actual Lead Time of an Order or Operation
LT_{target}	Target Lead Time of an Order or Operation

m	Total Number of Data Points
MAE	Mean Absolute Error
MCC	Matthews Correlation Coefficient
$m_{left/right}$	Number of Data Points in the Left/Eight Subset
NASD	No Assembly Start Delayer
P	Precision
PT	Processing Time
R	Recall
R^2	Coefficient of Determination
RMSE	Root Mean Squared Error
SS_{res}	Sum of Squares of Residuals
SS_{tot}	Total Sum of Squares
ST	Supply Time
SV	Schedule Variance
TAE	Time of Assembly End
TAS	Time of Assembly Start
TFPP	Time of Processing End of the First Supply Order
TLBV	Time of Processing End of the Last Supply Order
TN	True Negative
TP	True Positive
TPE	Time of Processing End
TPEP	Time of Processing End of Predecessor
TSD	Target Start Date
TSS	Time of Start of Setup
TT	Transition Time
VAO	Value of a Assembly Order
VSO	Value of a Supply Order
WT	Waiting Time
$y_{i,actual}$	Predicted Value
$y_{i,true}$	Actual Value

List of Figures

List of Tables

Introduction

Industrially manufactured products often consist of a large number of components sourced or produced using different manufacturing processes. This characteristic is particularly noticeable in the products of machinery manufacturers, whose products typically consist of a large number of components designed to meet specific customer requirements to provide a customized solution for each customer [1, 2]. In the globalized and internationalized procurement and sales markets, logistics performance, such as high adherence to delivery dates or short delivery and throughput times, is becoming increasingly important as a competitive factor. Particularly in Germany, a high-wage country, it is crucial for the success of companies to demonstrate excellent logistics performance to set themselves apart from international competitors, most of whom have more favorable cost structures [3–5]. Reliable logistics performance results in meeting promised delivery dates. If delivery dates are not met, the relationship with the customer suffers. In the worst case, a late delivery may cause a customer to choose a more reliable supplier for his next order. Conversely, high delivery reliability fosters high customer loyalty. As a result, on-time delivery has a positive impact on a company's profits and growth.

Studies in the engineering sector identify the main causes of poor logistics performance as high number of interfaces, complex order processing, long waiting times, poor planning and control, and unclear and inconsistent priorities [6, 7]. Particularly at points in the value chain such as assembly—where a large number of different components converge—the effects of disruptions along the entire process are noticeable [8].

© The Author(s), under exclusive license to Springer Fachmedien Wiesbaden
GmbH, part of Springer Nature 2024
F. Steinberg, *Machine Learning-based Prediction of Missing Parts for Assembly*,
Findings from Production Management Research ,
https://doi.org/10.1007/978-3-658-45033-5_1

1.1   In-time Supply of Goods for Assembly as a Success Factor

Swift delivery and reduced delivery times hinge on the punctual assembly of products. For an assembly process to stay on schedule, it is essential that all requisite parts are available when needed. Generally, the parts needed for assembly at machine manufacturing companies can be categorized into three types: components made within the company's production facilities; components procured on an order-related basis; and components procured generally and dispersed via a warehouse, such as typical C-parts [4, 9].

To ensure prompt delivery and prevent the delay of assembly initiation, it's crucial to schedule the delivery times and manufacturing processes. Within the production environment, calculating order completion dates using systems like Enterprise Resource Planning (ERP), Manufacturing Execution Systems (MES), and Advanced Planning and Scheduling (APS) is a commonly adopted approach for planning and control [10, 11]. These systems create the production schedule by taking into account available manufacturing equipment, technical constraints, due dates, and system status [12–14]. Nonetheless, even with these systems in place, disruptions like machine malfunctions, material shortages, personnel deficits, or inadequate employee skills can trigger production delays. To mitigate such disruptions, transparency of the assembly system's behavior is pivotal, and the use of data-based assembly models is advised.

Models like the assembly flow diagram or the material supply diagram [9, 10], well-established for the task of modeling the material supply process for assembly, have proven effective in pin-pointing general issues, such as a poor assembly supply situation in specific assembly areas. Although these general issues offer insights, they typically lead to the implementation of broad measures, like adjusting to more realistic lead times. However, in a machinery manufacturer's daily operations, it is beneficial to derive case-specific measures in addition to general ones. These case-specific measures are designed to accelerate production or procurement orders individually, ensuring parts are supplied in a timely manner for assembly. This underlines the need for more sophisticated prediction methods that can support the identification and execution of such case-specific measures. Machine learning (ML), as a promising Artificial Intelligence (AI) technology, could potentially provide the advanced predictive capabilities needed.

ML-based predictive models have seen increased use in two research areas of production management: production planning and control, and procurement. In the realm of production planning and control, existing models can predict lead times of individual production orders based on historical data (see, for example,

[15–18]). However, there's a notable gap concerning ML models that can predict potential individual missing parts for assembly. Furthermore, current lead time prediction models primarily rely on order and machine data and lack the use of feedback data from production data acquisition such as the a complete work status and material data such as dimensions of the produced part, indicating a potential area for improvement [19].

In procurement processes ML-based predictive models are already currently employed to forecast delivery delays of externally sourced products (see, for example, [20, 21]). However, these approaches face limitations, especially in low-volume, high-variety production environments, common in machinery manufacturing. Also, the existing models predict delivery delays only after the order is made and not at earlier times in the process such as the creation of an order request, hindering proactive responses. Further, they only use classification algorithms capable to predict if a purchasing order might be late or not, but they lack the capability to predict the severity a potential lateness, as regression algorithms could. Hence, there is a clear demand for more adaptable and proactive prediction models in this domain.

In conclusion, there is a distinct need for advanced prediction methods to ensure the on-time availability of assembly components. By addressing the limitations in current prediction methods and leveraging the potential of material data, we can enhance the accuracy and efficacy of models predicting missing parts for assembly. Further, these limitations lead to the main research question of this thesis: 'How should prediction models be designed and what data should be used within these models to sufficiently predict missing parts for the assembly?' The purpose of this dissertation is to show my findings on ML models that can predict the delay in assembly initiation due to the absence of components from internal production and external procurement. My research will evaluate which ML algorithm has the best predictive capability for missing in-house components. Furthermore, we'll measure the impact of material data on prediction quality. For the missing components acquired through procurement, we will examine if regression algorithms can accurately predict assembly initiation delays, thereby improving prediction quality.

The thesis is structured as follows: After defining the scope and objectives of the thesis as well as deriving the guiding research questions, section two summarizes the basics of the manufacturing process of a machinery manufacturer. Sections three to six cover four peer reviewed publications and thus represent the cumulative part of this thesis. The four publications are reproduced with standardized formatting and citation style. Finally, a critical reflection and summary is given in the last two section.

1.2   Research Methodology

According to CARNAP Sciences can be classified into formal and real sciences
[22]. The formal sciences, such as logic and mathematics, are concerned with
constructing sign systems governed by well-defined rules for their application.
On the other hand, the real sciences focus on empirically describing, explaining,
and designing observable aspects of reality. Within the real sciences, there are the
two categories: basic sciences and applied sciences [23]. Basic sciences, particularly the natural sciences, seek to explain theoretical relationships and enhance
our comprehension of the natural world. In contrast, applied sciences deal with
human decision-making processes. This includes decisions in production planning and control, as well as procurement, where humans determine factors like
task scheduling, machine allocation, and supplier and material choices. As such,
this study falls under the realm of applied sciences, necessitating a research
methodology appropriate for this scientific domain. In addition to the research
methodological classification, it is crucial to consider the underlying research
domain. This work addresses issues within the field of production sciences—
specifically, the minimization of missing parts before assembly commences—by
combining solution methods from the domain of computer science—employing
ML-based predictive models. As a result, this study lies at the intersection of production and computer science. Consequently, the applied research methodology
must encompass both research domains.

The research methodology of this work is following the Design Science
Research (DSR) methodology, which traces its roots to engineering and the
science of artificial intelligence [24]. DSR aims to expand human knowledge by
creating innovative artifacts to address real-world problems [25, 26]. Hevner's
DSR Framework [26] is an established approach for understanding, conducting, and evaluating DSR (see Fig. 1.1). The framework comprises three central
elements: Environment, Knowledge Base, and Design: The *Environment* encompasses the problem space where phenomena of interest reside, comprising people,
organizations, and existing or planned technologies. It involves defining goals,
tasks, problems, and opportunities perceived by stakeholders within the organization. The comprehensive understanding of the environment allows researchers
to identify specific research needs that demand resolution, forming the basis
of the research problem. This research problem is consequently dependent on
the respective real use case. Within this cumulative dissertation, three different
use cases in manufacturing companies are considered. Thus, the examination
of the environment is done individually for each use case in Chap. 4–6. The

Knowledge Base plays a pivotal role in addressing the research problem. It consists of two essential components: Foundations and Methodologies. Foundations draw from prior research and reference disciplines, offering theories, frameworks, instruments, constructs, models, methods, and instantiations. Methodologies, on the other hand, provide guidelines for the evaluation phase, ensuring rigor in the research process by appropriately applying existing knowledge. Analyzing the academic knowledge base determines the extent of readily available design knowledge to solve the problem of interest. The knowledge base for our use cases is defined in Chap. 2 and 3; case specific knowledge is further added within the Chap. 4–6. The core of DSR revolves around *design* activities, where innovative solutions are sought to address the research problem. Here, innovative solutions are created building upon and extending existing design knowledge. This iterative design process involves building and evaluating activities, refining solutions to achieve optimal outcomes. The design of the solutions is also use-case specific and is performed in Chap. 4–6.

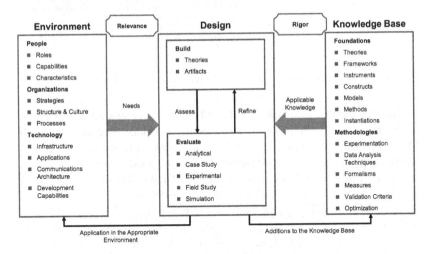

Fig. 1.1 Design science research framework [25, 26]

1.3 Objectives and Contribution of the Four Publication within this Cumulative Dissertation

Four research publications have been drawn upon to accomplish objectives of this thesis. The details of the publications are visually summarized in Fig. 1.2.

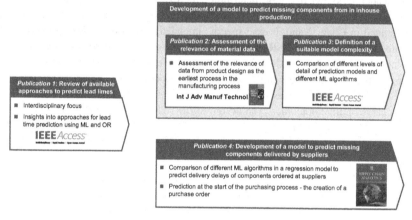

Fig. 1.2 Overview of the four publication within this cummulative dissertation

The first research paper (discussed in Chap. 3) is a comprehensive literature review that examines current methodologies in Operational Research (OR) and ML for predicting lead time, particularly in engineer-to-order production. As data and the applied algorithm significantly influence the model's effectiveness, this review aims to identify key data categories and OR/ML algorithms utilized for lead time prediction. Our research questions include: 'What is the state-of-the-art in direct lead time prediction for manufacturing companies, and what data, methods, or algorithms are being used?' and 'How does existing literature guide future research in direct lead time prediction?' Our systematic analysis reveals that ML is a growing research area. Mainly, order data and system status are the two data classes used for prediction. The use of material data (e.g., from PLM or ERP sources) and feedback data (e.g. from PDA) in complex models is notably minimal. Artificial Neural Networks and tree-based regression models show great promise in complex models considering material data and feedback data.

Considering that product design is a key process in machine manufacturing and components are often customized, it seems logical to assume that material data influences model quality. To verify this assumption, we conducted a case study at a machine manufacturing company and asked the following research questions: 'What effect does the use of material data have on the model quality of a model predicting assembly start delayers?' Within the case study, approach was to systematically compare the model quality of models with material data against those without. In detail, twelve different prediction models were developed to classify components as either *assembly start delayers* or *no assembly start delayers*. These models varied in their usage of material data and in the ML algorithm applied. The aim was to determine how the use of material data impacts the quality of a model predicting assembly start delay, while comparing various ML algorithms. The comparison verified that incorporating material data in models predicting assembly start delays enhances the model quality.

The third publication investigated the detail level of a model used for predicting assembly initiation delays in a company's own production. The level of detail can significantly influence model quality. Typically, increasing the detail level leads to improved model accuracy, though with a diminishing effect. Hence, the third paper's research question is: 'How does the detail level of a model affect its quality in predicting assembly start delays?' A total of 24 ML models were created to answer this question, each differing in their level of detail and applied ML algorithm but sharing the common objective of predicting assembly initiation delays. The comparison revealed that a binary classification model with surprisingly the least detail level was the best approach.

The fourth publication focuses on the procurement process. Based on our findings on internal production and our knowledge on the limitations of previous approaches predicting procurement delays, we formed three research questions. First, 'Can regression algorithms predict delivery delays of orders placed with low-volume, high-variety manufacturers?' Second, 'How does the prediction time[1], impact model quality?' And third, 'Is dimensionality a problem when developing a regression model predicting delivery delays of orders placed with low-volume, high-variety manufacturers?' A case study was carried out at a machine manufacturing company to answer these questions. We compared different ML models using various algorithms and successfully developed an ML-based regression model for predicting delivery delays of orders placed with suppliers in the engineering industry. Interestingly, we discovered that predicting delays at the early stages of the purchasing process can be effectively done using information available right after the internal order request is created and thus, even

---

[1] The prediction time referrers to the specific moment in a process the prediction model is executed and thus, the prediction is generated.

before the purchasing process starts. Also, there was no necessity to reduce the dimensionality of high-dimensional input features, debunking the notion of the 'curse of dimensionality' in this context.

To conclude, this research offers insights into the prediction of assembly start delays caused by missing parts, both in-house and from procurement. It brings into focus the importance of machine learning algorithms, the use of relevant data, and the level of model detail in enhancing prediction accuracy. Further, it emphasizes the potential of regression models in predicting delivery delays in the procurement process, setting the groundwork for future research in this area. Our findings underscore the applicability of ML models in improving operational efficiency in engineering and manufacturing industries.

Theoretical Background for the Prediction of Missing Parts for Assembly

2

To facilitate a comprehensive evaluation of the novelty and relevance of the four publications in this cumulative dissertation, the following sections present a detailed overview of the manufacturing processes involved. These processes are essential for ensuring the on-time supply of components required for assembly commencement (cf. sect. 2.1). Furthermore, the dissertation explores the planning and control processes associated with manufacturing, which significantly impact the punctual delivery of components for assembly initiation (cf. sect. 2.2).

2.1 Fundamentals and Challenges of a Manufacturing Process

The value chain of manufacturing companies can be differentiated into four ideal-typical types: Engineer-to-order production, make-to-order production, assemble-to-order production, and make-to-stock production [5, 27]: For an engineer-to-order manufacturer, a single customer order initiates the value chain and creates an individual demand. This demand is characterized by products to be produced according to individual customer specifications. Thus, the design of a product is based on the individual requirements of customers. Each customer order, therefore, represents a new, customized or variant design [28]. Due to the considerable complexity of the products, engineer-to-order manufacturers focus on their core competencies in development, design and assembly. This circumstance entails a high degree of outsourcing of various production steps to other companies and a large proportion of externally sourced components.

© The Author(s), under exclusive license to Springer Fachmedien Wiesbaden GmbH, part of Springer Nature 2024
F. Steinberg, *Machine Learning-based Prediction of Missing Parts for Assembly*, Findings from Production Management Research , https://doi.org/10.1007/978-3-658-45033-5_2

For the remaining internal components the underlying manufacturing process r consists of the process steps design, production planning, procurement of raw material, production of components and assembly (see Fig. 2.1) [5, 29, 30]. To obtain a basic understanding of the respective process steps along the manufacturing process, the steps are specified in Sects. 2.1.1 to 2.1.5.

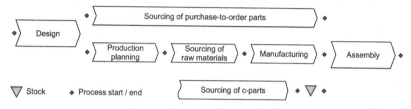

Fig. 2.1 Manufacturing process of an engineer-to-order manufacturer (according to [30–32])

2.1.1 Design

Within the manufacturing process, the design process encompasses all activities from the product idea to the generation of production documents (drawings, bill of materials, etc.), whereby the product characteristics must meet the customer's requirements [33, 34]. According to [34, 35], the design process can be divided into four steps (cf. Fig. 2.2):

1. The design process starts with the clarification of the task, e.g., in the form of a specification sheet. For this purpose, the designer obtains information about the concrete product requirements and framework conditions. The result is an informative specification in a list of requirements differentiated according to the mandatory requirements and wishes of the customer.
2. In the second step, a solution is defined in principle based on the requirements list. For this purpose, the designer sets up the functional structures with main and sub-functions and formulates the physical operating principles for implementing the individual sub-functions. The definition of a specific combination of the various operating principles results in a functional structure for fulfilling the main functions.
3. Based on the functional structure, the clear and complete design structure is created in the design phase according to technical and economic aspects.

For this purpose, the designer first prepares preliminary designs and estimates their respective advantages and disadvantages. Finally, he determines the final overall design by combining the best partial solutions of the designs.

4. In the final phase of design, the designer supplements the previously defined design structure with specifications for the shape, dimensioning and surface quality of all individual parts. In addition, he determines all materials, checks the manufacturing and assembly possibilities and fixes his decisions in binding documents such as parts lists and drawings. The result is thus the technical manufacturing specification of the solution.

Fig. 2.2 Design process (according to [10])

As a result, the specifications of the component, and thus the requirements for the manufacturing process, are determined in the design process, where the defined components are either manufactured in-house or purchased from suppliers. It is obvious that information from the design process, such as the dimensions or materials of the components is a valuable source for models predicting missing parts at the start of assembly (cf. Chap. 4).

2.1.2 Production Planning

Based on the bill of materials and drawings created in the design process, all planning tasks are carried out in the production planning process to ensure that a product is manufactured according to the product requirements [10, 30, 36]. These tasks include material planning, work plan creation and NC programming:

- Material planning: Material planning determines the type and dimensions of the required raw parts. Technological criteria such as machining properties, shape and surface as well as procurement times must be considered. The time criteria pose major challenges, especially for casting and forging components, which are frequently subject to tight deadlines.
- Work plan creation: A work plan differentiates a production order into individual operations and assigns for example the required materials, machines, and standard times.

- NC programming: NC programming then adds the appropriate programs for the machines to enable the production process.

Production planning serves as a crucial bridge between the conceptualization stages—design and procurement of raw materials—and the actual fabrication of components. This planning phase generates vital documents and information that guide the internal manufacturing process. The work plan assumes significant importance. It outlines the machines to be used and sequences the processing of components, complete with estimates of the time required for each operation. This data is indispensable for order scheduling and determining the requisite production capacity. Therefore, it is vital to incorporate the insights garnered from the production planning process into a predictive model. This model will help forecast any potential shortages of parts at the onset of assembly, thereby minimizing disruptions and enhancing efficiency in the production process (cf. Chaps. 4 and 5).

2.1.3 Procurement

Procurement includes all necessary activities that ensure the availability of all materials required but not produced in-house. It is important to ensure that the materials are available in the right quantity and quality, at the right time and at the lowest possible cost.

According to ARNOLD [37] the procurement process can be divided into seven typical subprocesses (cf. Fig. 2.3):

1. After receiving a demand notification, the procurement process starts with an inquiry to the supplier. This inquiry should contain all technical specifications and business conditions. These include, for example, type of material, quantities, drawings, descriptions, method of payment and delivery date. The number of offers to be received by suppliers depends on the procurement volume.
2. As soon as the suppliers return offers, a selection process starts to determine the optimum offer. To do this, it is first checked whether the content of the offer matches the inquiry. Then the offers are compared according to criteria such as price, delivery time, quality, supplier location or trust in the supplier.
3. If only one offer turns out to be optimal, an order is placed at the corresponding supplier directly. However, if several comparable offers exist, further procurement negotiations will be conducted with the suppliers before placing the order.

4. As soon as all negotiations are finished, one supplier is selected.
5. With placing the order at the supplier legally binding contract is created. This contract contains the final specifications from the offer, inquiry and procurement negotiation.
6. The supplier sends an order confirmation after receipt of the order and, if necessary, further confirmations (e.g., delivery notifications).
7. As soon as the supplier completes the order, he delivers the material to the incoming goods area of the ordering company. There, the procurement process ends with a delivery note check, quantity and quality check, and the storage as well as payment of the goods.

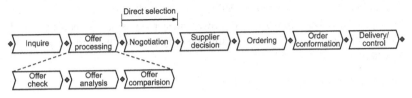

Fig. 2.3 Typical sub-processes of procurement (according to [37])

In practice, different procurement models are established depending upon the procurement release, the kind and the place of warehousing, and the property transition between supplier and customer [10, 38–40]. In the following, the procurement models typical for an engineer-to-order manufacturer (cf. Fig. 2.4) are explained according to definitions in [10, 38–40]:

- In stock procurement, material inventories are held in a targeted manner, thus ensuring the security of supply for the subsequent process steps. Orders are typically placed at suppliers based on the available stock in the warehouse.
- In the case of individual procurement, procurement always takes place in relation to a specific (customer) order. The required material is delivered directly to the place of further processing without intermediate storage.

Thus, the procurement process includes all activities to procure raw materials for in-house production and to procure engineer-to-order parts for assembly. As a result, the procurement process directly influences the availability of materials at the start of assembly. On the one hand, the procurement process can be responsible for the fact that the raw materials required for the in-house processes are

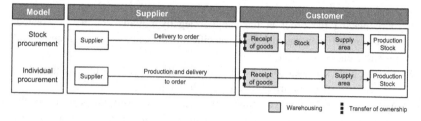

Fig. 2.4 Selected procurement models (according to [39])

not available in time. As a result, components from in-house production may not be available in time for assembly. On the other hand, the procurement process has a direct impact on the timely delivery of externally purchased components. Consequently, information from the procurement process should be considered when developing a model for predicting missing parts at the start of assembly (cf. Chap. 6).

2.1.4 Component manufacturing

In component manufacturing, the individual components for assembly are manufactured in a single- or multi-stage process [41]. According to [5], this process can be divided into five subprocesses (see Fig. 2.5):

1. After a production order has been created, it is scheduled with time buffers and adjusted to the available capacity. Subsequently, a workshop program is created in which it is described at which times, and on which machines the order is to be executed.
2. Based on the workshop program, a precise check of the availability of material and capacities is carried out again before the start of production.
3. If the check is positive, the order is released. The release also includes the handover of production documents (design drawing, work plan, parts list, etc.) to the production personnel.
4. As soon as the production order is released and the production documents have been handed over, the order is executed and controlled. This sub-step represents the physical manufacturing process of the product.
5. After the completion of the execution of the order, the execution is confirmed to close it.

Fig. 2.5 Component manufacturing (according to [5])

The execution of an order can be divided into individual operations, which in turn are subdivided into the sub steps waiting after processing (of the predecessor), transport, waiting before processing, setup, and processing [11]. Accordingly, an operation begins with the completion of the respective predecessor operation (see Fig. 2.6)

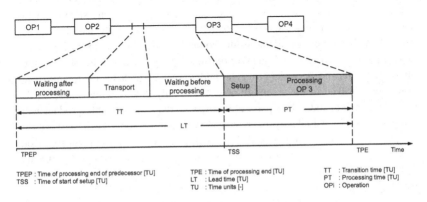

Fig. 2.6 Manufacturing throughput element [11]

According to [11], the period from the end of processing of the predecessor to the end of processing of the operation under consideration is the lead time of the operation. The execution time is made up of the time portions of the subprocesses setup and processing. The difference between lead time and execution time constitutes the transition time.

Considering information from this process is obviously valuable when setting up prediction models to predict missing parts for assembly. Within this cumulative dissertation, publication I (cf. Chap. 3) is giving an overview of the state of the art in lead time prediction and publications II and III (cf. Chaps. 4 and 4) are focusing on ML models predicting missing parts from in-house manufacturing process.

2.1.5 Assembly

In addition to the processes upstream of assembly, which determine the availability of parts at the start of assembly, it is also important to understand the assembly process for a comprehensive understanding of the process. In contrast to Fig. 2.1, the assembly process is preceded by a supply process, which functions as an interface between assembly and the processes upstream of assembly [32, 42]. Material supply can be understood as a central process in which many upstream processes converge before they enter the assembly. Consequently, the supply of materials to the assembly represents a convergence point. A convergence point in the manufacturing process offers a high potential for delays since assembly can only be executed if all the required components are supplied without defects and on time. According to REFA, material supply has the task of "providing the material available in the company for use in the execution of the task in the required type and quantity on time " [43]. Assembly follows the supply of materials and, according to VDI Guideline 2860, includes the totality of all processes involved in the assembly of geometrically defined bodies [44]. In the assembly, several individual parts are joined so that a marketable product with a higher level of complexity is created [45]. Since assembly is the last step in the production value chain, it is also referred to as the collection point for errors and faults in the previous production processes [42, 46]. According to [10], the assembly process can be single or multi-stage depending on the product structure (see Fig. 2.7):

- In a single-stage process, all components manufactured by the upstream processes are assembled in one assembly operation.
- In a multi-stage process, individual components are first assembled into modules. These modules are then combined with other components or modules at a higher level, also in modules or the final assembly. With the help of modules, the assembly process can be parallelized and thus accelerated.

With the assembly flow element of SCHMIDT [9] (see Fig. 2.8), a model for logistics within assembly exists: In this model, the consideration of an assembly process starts with the completion of the respective predecessor process. It considers the convergence of several material flows and deducts the time portions of the sub-processes of an assembly operation on the time axis. In detail, these time elements are waiting after the predecessor process, transport, waiting before assembly, setup and assembly. For example, in Fig. 2.8, three predecessor processes are considered. After finishing the last predecessor process (TLBV$_i$), the

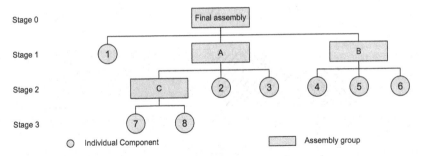

Fig. 2.7 Product structure in a multi-stage assembly [47]

assembly process is considered as completed. According to the assembly flow element, the assembly lead time consists of two-time components, the execution time and the transition time: The execution time includes the assembly activity and the setup process. The transition time includes the activities 'lying after predecessor process', 'transporting' as well as 'lying before assembly', and is made up of the time components supply time and waiting time. The supply time describes the time in which the predecessor processes of an assembly operation are completed, and the respective components arrive at the assembly system. It, therefore, represents the difference between the processing end date of the last predecessor process and the processing end date of the first predecessor process. The completion of the predecessor process thus represents the end of the supply time and the start of the waiting time, which extends to the assembly start.

In contrast to the assembly flow element, which describes a single assembly process on an assembly system, the assembly flow diagram depicts several assembly processes on an assembly system [9]. To create the assembly flow diagram, all assembly flow elements under consideration at the respective assembly system are first sorted according to their respective outflow date and plotted across the time axis (see Fig. 2.9a). The individual assembly flow diagrams are then separated into their supply processes as well as the assembly process, and four curves are then created from the temporal variables (inflow, completion, start of assembly and outflow). The outflow dates of the assembly flow elements, weighted with their respective values, are plotted cumulatively over time forming the outflow curve. The same applies to the inflow, completion and assembly start curves. The procedure is visualized exemplarily for the assembly pass element marked with M (see Fig. 2.9b). With the assembly flow diagram, in addition to the usual indicators in a flow chart (e.g., stock or capacity), the disturbed and completed

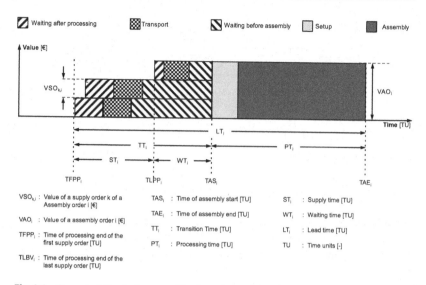

Fig. 2.8 Characteristics in the assembly throughput element [9]

stock can be seen. The disturbed stock results from the deviation between the inbound and the completion curve (see Fig. 2.9c: ID). The completed stock, on the other hand, results from the deviation between the completion and outflow curve (see Fig. 2.9c: IC).

With the assembly flow chart, it is possible to evaluate the logistic state of an assembly system based on four curves—inflow curve, outflow curve, completion curve and assembly start curve—as well as various key figures such as the stock or capacity. As a further development of the assembly flow diagram and taking into account the completion curves of KETTNER [48], SCHMIDT [9] developed the assembly supply diagram, which serves as a model for the analysis of material supply for assembly. The starting point for the creation of the assembly supply diagram is a target/actual comparison in the assembly flow chart (see Fig. 2.10a). Subsequently, the assembly flow elements are normalized to their due or target dates (see Fig. 2.10b). The third step is to sort all supply orders according to their respective inflow date deviation and to plot them cumulatively in a curve, the so-called normalized inflow curve. The dashed lines indicate this as an example for three of the incoming supply orders. Afterwards, the completed assembly orders are sorted according to the schedule deviation of their respective last finished supply order and plotted in the so-called completion curve (see

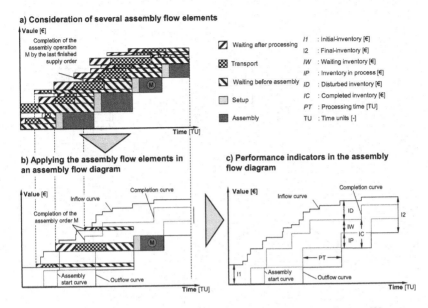

Fig. 2.9 Derivation and performance indications of the assembly flow diagram [9]

Fig. 2.10c). The discrete representation determined in this way can be transferred to an idealized representation in which key figures such as the value of assembly operations completed at the demand date, or the value of supply orders not yet provided at the demand date can be seen (see Fig. 2.10d). The idealized form of the assembly supply diagram is based on the approximations for the inflow and completion curve developed by BECK [32]. The inflow curve is determined utilizing distribution functions and the total inflow values of the supply orders.

Overall, the assembly supply diagram thus enables a past-related evaluation of material availability at the start of the assembly. It represents the actual state of the supply orders of various assembly orders after their completion. In the assembly supply diagram shown as an example, there are supply orders with both positive and negative schedule deviations in relation to the due date (see Fig. 2.10d). There are two basic approaches to improving schedule variance, which can be combined (see Fig. 2.11) [28]:

- Timeliness (see Fig. 2.11 top) aims at reducing the proportion of those supply orders with positive schedule deviation (delay). Measures to achieve this

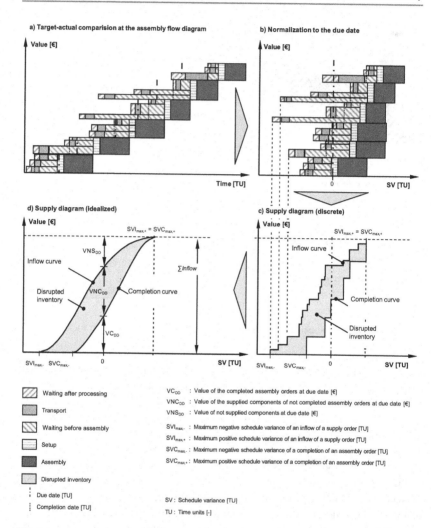

Fig. 2.10 Derivation of the assembly supply diagram based on the assembly flow diagram [9, 32]

include realistic planned lead times, realistic procurement times, deadline-oriented sequencing and the scheduling of buffer times before assembly. However, in the case of a buffer, care must be taken to avoid triggering the so-called error circuit of production control [49, 50]. Successful application of these measures results in steeper curves in the readiness diagram.

- Simultaneity (see. Fig. 2.11 middle) aims to synchronize the supply orders of an assembly operation with each other so that the assembly order is completed in as narrow a time window as possible. This is enabled by a deadline-oriented order release and procurement as well as sequencing and procurement according to order structure and deadline situation. These measures result in a closer alignment of the two curves in the assembly supply diagram.
- A combination of both approaches results in a punctual supply of materials for assembly (see Fig. 2.11 bottom). Further measures include increasing the capacity flexibility of production and shortening procurement times. The result of punctual supply is a reduction in incomplete stocks as well as an increase in adherence to delivery dates.

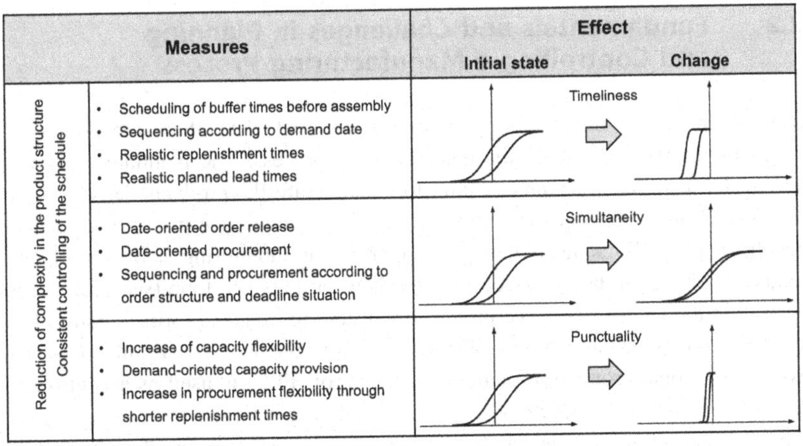

Measures		Effect	
		Initial state	Change
Reduction of complexity in the product structure / Consistent controlling of the schedule	• Scheduling of buffer times before assembly • Sequencing according to demand date • Realistic replenishment times • Realistic planned lead times	Timeliness	
	• Date-oriented order release • Date-oriented procurement • Sequencing and procurement according to order structure and deadline situation	Simultaneity	
	• Increase of capacity flexibility • Demand-oriented capacity provision • Increase in procurement flexibility through shorter replenishment times	Punctuality	

Fig. 2.11　Measures to improve the supply of materials for assembly [51]

However, the assembly supply diagram is primarily designed to analyze historical data and general issues such as an overall poor assembly supply situation in individual assembly areas. Furthermore, the measures that can be derived from the assembly supply diagram to improve material supply to the assembly are only general measures such as defining realistic lead times or adjusting lead times. In the daily business of a machine manufacturer, however, it can be helpful to derive not only general measures, but also individual case-related measures to accelerate individual production or procurement orders in order to ensure the on-time supply of components for assembly. For these case-related measures, it is especially helpful if they can be taken as early as possible in order to have enough time to implement the measure. Therefore, this dissertation examines how to predict potential missing parts at the beginning of assembly in order to derive individual case-related measures.

Overall, the process steps described in this section serve to provide a basic understanding of the respective processes within a machine builder's value chain. They also illustrate the complexity of the manufacturing process and the resulting challenges that a machine builder must overcome to complete its final products on time and to customer specifications. Each process has a specific role to play and can cause delays in the overall manufacturing process. Therefore, all of the processes described in this section must be considered when predicting missing parts for the start of assembly.

2.2   Fundamentals and Challenges in Planning and Controlling a Manufacturing Process

In addition to the manufacturing processes (cf. sect. 2.1), there are processes for planning and controlling the manufacturing process. The planning and control of the manufacturing process aim to manufacture the products derived from customer requirements at the appropriate quality, competitive prices, and agreed deadlines [10]. To achieve this goal, appropriate monitoring and control are required, which can be described with the help of a control loop (see Fig. 2.12).

Within the control loop, the production system (execution) forms a controlled system that can deviate from its planned behaviour due to external influences. To avoid deviations, production planning and control (PPC) is used as a controller. The PPC plans the production program considering a target, the needs of customers or sales and the current state of the controlled system. In addition, within the control loop, three types of status values—target, plan, and actual—exist, which serve as logistical planning variables for the controller. H-H. WIENDAHL [4] defines them as follows:

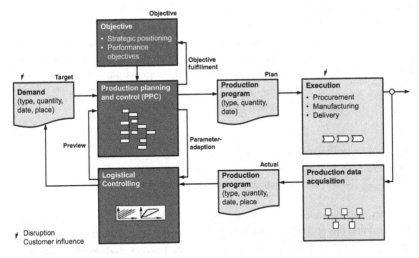

Fig. 2.12   Production management control loop [10]

- The target variable describes the desired logistical behaviour. In the sense of control loop-based action, it serves as a reference. Changes to the target variable can only be made in agreement with the customer.
- The actual variable represents the actual logistical behaviour determined by production data acquisition and is, therefore, a past-related variable.
- The plan variable describes the expected logistic behaviour under consideration of the existing boundary conditions and thus represents a forecasted actual variable. Occurring disruptions can cause deviations from the plan from the target.

In a production environment, the plan variables are determined by the production schedule considering the available production capacity, technical restrictions, due dates (target variable) and the system status (actual variable) [13, 14, 52]. The job sequence is determined according to certain rules to calculate the start and end dates of the jobs at the workstations [53]. Of course, disruptions can occur that lead to a deviation from the schedule. In this case, a rescheduling is performed to update the schedule according to the new situation [14].

Based on the logistical planning variables, a start and finished date deviation, as well as a lead time deviation, can be calculated for each production order and operation [4, 28]. In the following, the individual deviations are described at the level of an operation. The definitions apply analogously to a production order. The finished date deviation is defined as the deviation between the actual and the target finished date (due date) of the operation under consideration. A positive date deviation, therefore, indicates a too late completion, a negative date deviation, on the other hand, indicates a too early completion.

$$DD_{finished} = DOF_{actual} - DOF_{target} \qquad (2.1)$$

$DD_{finished}$ = Finished date deviation [TU]
DOF_{actual} = Actual finished date of an order or operation [TU]
DOF_{target} = *Target* finished date of an order or operation [TU]

The start date deviation, on the other hand, is defined as the deviation between the actual and target start dates of the operation under consideration. Accordingly, a positive start date deviation indicates a too late process start, while a negative start date deviation indicates a too early process start.

$$DD_{start} = DOS_{actual} - DOS_{target} \qquad (2.2)$$

DD_{start} = Start date deviation [TU]
$DOS_{actuals}$ = Actual start date of an order or operation [TU]
DOS_{target} = Target start date of an order or operation [TU]

The lead time deviation—also called relative date deviation—is defined as the deviation between the actual lead time and the target lead time of the operation under consideration. The lead time is calculated equally for target and actual variables as the difference between the start and end of a process step. A positive value corresponds to a longer lead time than the target lead time and a negative value to a shorter lead time.

$$\begin{aligned} DD_{rel} &= LT_{actual} - LT_{target} \\ &= (DOF_{actual} - DOS_{actual}) - (DOF_{target} - DOS_{target}) \end{aligned} \qquad (2.3)$$

DD_{rel} = Relative date deviation [TU]
$LT_{actuals}$ = Actual lead time of an order or operation [TU]

LT_{target} = Target lead time of an order or operation [TU]

In summary, the aspects described serve as a basic understanding and show the complexity of planning and controlling the manufacturing process. All information from the manufacturing processes of the individual components converges in the planning and control process. Here, orders are scheduled in such a way that there are as less as possible delays compared to the target date, delays resulting from the schedule are indicated and actual deviations from the schedule are calculated based on feedback data from the execution of the orders. One of the fundamental rules to schedule orders is to determine the job´s waiting time depending on the machine´s utilization [54]. Here, performance curves play a key role [55]. The performance curves, also called operating curve [56] or characteristic curve [57], can be generally understood as a tool to model performance indicators of a workstation´s productivity considering functional relationships between logistic parameters such as lead times, throughput and stock [54]. Consequently, the derivation of the schedule and thus of the potential deviations within the schedule is based on defined parameters within the manufacturing process. Information about past behaviour of the manufacturing system as well as information from processes upstream of production, such as information about the design of a product, are not considered in the performance curves and thus, are typically not considered when identifying potential deviation within the schedule. The focus of this dissertation is to use this additional information from other processes and from the past to improve the prediction of missing parts for assembly.

Publication I: Approaches for the Prediction of Lead Times in an Engineer to Order Environment—a Systematic Review

3

P. Burggräf, J. Wagner, B. Koke, and F. Steinberg, "Approaches for the prediction of lead times in an engineer to order environment—a systematic review," *IEEE Access*, vol. 8, pp. 142434–142445, 2020, https://doi.org/10.1109/ACCESS.2020.3010050.

3.1 Abstract

The interest of manufacturing companies in a sufficient prediction of lead times is continuously increasing—especially in engineer to order environments with typically a large number of individual parts and complex production processes. A multitude of approaches have been proposed in the literature for predicting lead times considering different data and methods or algorithms from operations research (OR) and machine learning (ML). In order to provide guidance at setting up prediction models and developing new approaches, a systematic review of the available approaches for predicting lead times is presented in this paper. Forty-two publications were analyzed and synthetized: Based on a developed framework considering the used data class (e.g., product data or system status), the data origin (master data or real data) and the used method and algorithm from OR and ML, the publications are classified. Based on the classification, a descriptive analysis is performed to identify common approaches in the existing literature as well as implications for further research. One result is, that mostly order data and the status of the production system are used for predicting lead times whereas material data are used seldom. Additionally, ML approaches primarily use artificial neural networks and regression models for predicting lead

F. Steinberg, *Machine Learning-based Prediction of Missing Parts for Assembly*, Findings from Production Management Research , https://doi.org/10.1007/978-3-658-45033-5_3

times, while OR approaches use mainly combinatorial optimization or heuristics. Furthermore, with increasing model complexity the use of real data decreased. Thus, we identified as an implication for further research to set up a complex data model considering material data, which uses real data as data origin.

3.2   Section I: Introduction

Production companies are in a constant state of change. They are challenged to assert themselves in international markets. Growing demands for individualized products with increasing quality and decreasing prices bring logistics performance, such as high adherence to deadlines or short delivery and lead times, to the fore as a competitive factor [29, 58]. As a result, lead time is one of the key factors for meeting customer requirements [59]. By means of a valid prediction of the lead times, delivery dates can be determined at an early stage and deviations from schedule can be identified [60]. In contrast, an imprecise prediction of lead times can lead to delivery dates not being met, resulting in loss of customer confidence and consequential costs for late deliveries [12]. Particularly relevant is the prediction of lead times for mechanical and plant engineering, a typical example of an engineer to order process. In addition to production and assembly, here the lead time includes all upstream processes such as design, order planning or the purchasing process for raw materials and finished parts [5]. Furthermore, the products of a machine and plant manufacturer often consist of a large number of components that are designed individually to achieve a tailor-made solution for the respective customer [1, 2]. Consequently, the product characteristics defined in the design process represent a unique selling point for the companies.

A primary cause of not meeting due dates and extended lead times are the negative effects of disruptions [61, 62]. The occurring disruptions are manifold and include, for example, machine breakdowns, missing material, lack of personnel or insufficient employee qualification [63, 64]. However, a recent study found that the majority of disruptions in the assembly process occur repeatedly and are theoretically predictable [65]. If the occurrence of a disruption is known or predictable, this information should also be considered to predict lead times. Consequently, data containing that information about disruption and thus the causes of delays should be used for the prediction of lead times. The number of potential data classes, however, varies due to the large number of possible disturbances. In addition to the considered data, the methods and algorithms used for the prediction are relevant for the quality of the prediction [66, 67]. For the prediction of lead times, methods and algorithms from the field of operations research (OR) such as heuristics or combinatorics and from the field of machine learning (ML) such as neural networks or random forest can be applied [15–18]. Consequently, the question arises which data should be considered in the context of the forecast and which method or algorithm can be utilized. Due to the multitude of possibilities, the choice is not easy. A systematic review can help to achieve an overview of the existing methods and thus facilitate the selection for the user.

In their often cited survey CHENG AND GUPTA [68] investigated relationships between due dates, dispatching rules and completion times in static and dynamic job shops. ÖZTÜRK, KAYALIGIL ET AL. [69] comprehensively summarized the development of prediction models with a focus on dispatching rules and scheduling. LINGITZ, GALLINA ET AL. [70] focused on approaches with regression models to predict lead times. KARAOGLAN UND KARADEMIR [71] provided a comprehensive overview of the mathematical approaches used in the field of machine learning as well as the data classes considered. In all publications, however, only parts of the current state of the art are considered. In addition, it is not always possible to identify whether a systematic procedure was used to review the literature. Even after a comprehensive search, no review was found that systematically summarized both the state of the art of the methods and algorithms used and the data considered.

The aim of this paper is therefore to conduct a systematic literature review to answer the following research question: 'Which is the current state of the art in predicting lead times in engineer to order environments and which data and methods or algorithms are used?'. Additionally, we ask as second research question 'How does the existing literature contribute to future research on the prediction of lead times?' to identify implications for further research. In our study we follow the structure of VOM BROCKE ET AL. [72] supplemented by dedicated review concepts from other authors like a procedure model of MOHER, LIBERATI ET AL. [73] and a clustering approach of WEIßER, SAßMANNSHAUSEN ET. AL. [74]. Since we assume that the authors use different classes of data and methods or algorithms, we will develop a framework for the classification of the publications. Based on the classification, we will perform a descriptive analysis, which will then be used to identify focus topics in the existing literature as well as implications for further research.

Our paper is structured as follows. Section II first introduces the terms lead time and prediction. Section III elaborates the systematic literature review and details the applied methodological approach. In section IV a framework is derived as a result of the systemic review and a detailed analysis of the current state of the art in the body of literature is conducted. Based on this, the implications for further research are derived in section V. Finally, a summary is given in the last section.

3.3   Section II: Lead Time and Lead Time Prediction

According to the Business Dictionary [75] *lead time* is defined as the 'number of minutes, hours, or days that must be allowed for an operation or process, or must elapse before a desired action takes place'. A definition for the term lead time with focus on manufacturing processes is given by the Cambridge Business English Dictionary [76] and GUNASEKARAN, PATEL ET AL. [77] with the time that elapses between receiving a customer's order and the delivery of the goods or service to the customer. A more detailed definition for the manufacturing lead time is given by the Business Dictionary with the 'total time required to manufacture an item, including order preparation time, queue time, setup time, run time, move time, inspection time, and put-away time. For make-to-order products, it is the time taken from release of an order to production and shipment' [75]. WIENDAHL [28] and NYHUIS [11] divide an order into individual operations and differentiate accordingly between order lead time and operation lead time: The order lead time elapses between the start of the first operation and the end of the last operation. Each operation lead time is further divided into the interoperation and operation time. The interoperation time consists of the three components wait time after processing of the previous operation, time for transportation between previous and current workstations and another waiting time before processing on the current workstation. The operation time is divided into the setup time and the actual processing time. As it is well known, waiting times have a higher share in the lead time than the processing times [28, 78, 79].

In a production environment the job's lead times are determined by the production schedule considering the available production capacity, technical restrictions, due dates and the system status [12–14]. The job sequence is defined according to certain rules to calculate the start and end dates of the jobs at the work stations [53]. One of the fundamental rules is to determine the job's waiting time depending on the machine's utilization [54]. Here, performance curves play a key role [55]. The performance curves, also called operating curve [56] or characteristic curve [57], can be generally understood as a tool to model performance indicators of a workstation's productivity considering functional relationships between logistic parameters such as lead times, throughput and stock [54]. To determine the performance curves, several different methods are known, which are subdivided mainly into the two areas *approximation function* and *queuing theory*

[54, 55]. Within the area of *approximation functions* the main representative is a description of elementary relationships of flow processes based on the so-called "funnel model" and the flow diagram [11, 28, 80]. The funnel-model focuses on the representation of the performance-stock ratio and determines the capacity of a workstation as the upper performance limit. Here, the performance curve is defined as a so called C_{Norm}-function [11]. The area of *queuing theory* condenses approaches which are mainly based on the so-called Kingman equation [81], as well as their extensions to multi-operator systems and adaptations for practical use (see [55] and [82], and the references herein for further details). One exemplary extension of the Kingman's equation is given by the authors in [83], who approximated the curve by using a constant factor to replace the variability term in the Kingman's equation. The authors in [84], [85], [86] and [87] used this extension to quantify the productivity improvement of a semiconductor fabrication plant. Furthermore, historical data can be used in the determination of performance curves. WU AND MCGINNIS [88] for example used historical lead times in the determination of the performance curves and based on that calculated queueing times and subsequently lead times.

After determining the production schedule, of course, disruptions can occur that lead to a deviation from the schedule. In this case a rescheduling is performed to update the scheduled according to the new situation [14]. There are also approaches that consider potential disruptions during scheduling to get a more robust schedule [14]. LEON, WU ET AL. [89] for example analyze the effect of single disruptions for delaying a job and use a genetic algorithm that minimizes expected delays and lead times to find a robust schedule. TADAYONIRAD, SEIDGAR ET AL. [90] take unplanned machine breakdowns into account. Summarized in both scheduling and rescheduling the expected lead time is calculated based on the determined job sequence and available capacities.

Besides calculating the lead time based on a previous sequencing the lead time can also be predicted directly. In the past, a large number of approaches have been established for predicting lead times. CHENG AND GUPTA [68] performed an early literature review and investigated relationships between due dates, dispatching rules and lead times in static and dynamic job shops. Their focus was on a particular segment of scheduling research in which the due date assignment is of primary interest. They reviewed methods for calculating a job's due date based on a given job starting time and a predicted flow allowance, which is equal to a lead time. They differentiated between exogenous and endogenous methods [68]. In exogeneous methods, a job's lead time is set as a fixed and given attribute of a job before entering the production system. Examples are Constant (CON), where all jobs are given exactly the same lead time, and Random (RAN), where the lead time for a job is randomly assigned. In endogenous methods the job's lead time is predicted as the job is entering the production system considering job characteristics and shop status information. Examples for considering job characteristics are Total Work (TWK), where the lead time is predicted based on a jobs processing time and Number of Operations (NOP), where lead times are predicted based on the number of operations to be performed on the job. Examples for considering shop information are Jobs in Queue (JIQ), where the lead times are predicted based on the number auf jobs in a queue of the production system or Work in Queue (WIQ), which is similar to JIQ but utilizes the processing times instead of the number of jobs. Comparing the predicted lead times of exogenous and endogenous methods, the endogenous methods are generally superior [91]. Combining job and shop status has proven to be more effective [92, 93]. Further details on the methods and its performance are given by [92, 94, 95]. All approaches reviewed by CHENG AND GUPTA have in common that they use analytical techniques for the prediction of lead times that are typically found in in the field of OR. One of the most fundamental analytical approaches is Little's Law, which determines the average number of items in a queue of a stationary system based on the average arrival rate of items to that system and the average waiting time [96]. With the increasing development of ML, new data analytics methods for directly predicting lead times have emerged. In their study, BURGGRÄF, WAGNER ET AL. [97] have highlighted that scheduling and the prediction of lead times was traditionally one of the key research topics for ML in production. ÖZTÜRK, KAYALIGIL ET AL. [69] for example used a regression tree to predict lead times considering several attributes from shop status and job characteristics which outperforms the traditional TWK, ALENEZI, MOSES ET AL. [98] utilize a support vector machine and WANG AND JIANG [99] develop a deep neural network.

Concluding, there are two possible approaches to determine lead times: Firstly, indirect based on scheduling and approximating waiting times considering performance curves and secondly, by performing a direct prediction of lead times based on specific rules or historical data. To the best of our knowledge, no review article analyses the current status of available approaches for the direct prediction of lead times coming from both areas ML and OR. In the recent works the relevant state of the art is summarized. However, no systematic procedure is apparent.

3.4   Section III: Conducting the Review

A systematic review is a type of literature review based on systematic methods to reproducibly answer a specific research question by identifying all relevant studies and synthesizing findings qualitatively or quantitatively [100, 101]. It is designed to provide a complete, exhaustive, transparent and replicable summary of current stare of the art [102].

The methodology used in this review is following the procedure model of VOM BROCKE ET AL. which consists of five steps: (I) definition of review scope, (II) conceptualization of topic, (III) literature search, (IV) literature analysis and synthesis as well as (V) deduction of research agenda [72]. It is widely accepted within review theory [103] and not least it grants freedom of action for domain and process specific examinations.

3.4.1   Definition of Review Scope

The review scope was characterized according to the taxonomy of literature reviews by COOPER [104] (cf. Fig. 3.1). The research focus is on research outcomes and applications with the goal of knowledge integration using a conceptual structure. From a neutral perspective the review addresses specialized scholars considering all the relevant sources, but describing only a sample. So, the coverage is classified as exhaustive and selective.

The organization of prior research identifies a relationship between the considered data, algorithms and predicted lead times and serves to highlight the high multitude of possibilities to predict lead times (cf. Section II). The aim of this systematic literature review is consequently first to aggregate the latest state of the art for the prediction of lead times including used data and algorithms and second to develop an integrative framework for the further analysis and synthesis of the relevant publication. Here, we want to focus on the direct prediction of lead times only and leave out approaches focusing on scheduling, queueing theory or performance curves since these approaches rely on the determination of waiting or interoperation times and do not fully consider potential disruptions occurring during production process itself leading to an extension of the processing time. A direct prediction of lead times can include these disruptions as it considers always the complete lead time consisting of waiting and processing time instead of only a part of it. Furthermore, a direct prediction of lead times based on historical data is gaining new potentials with the enormous improvements in data

Characteristic	Categories			
1	Focus	Research outcomes	Research methods	Theories	Applications
2	Goal	Integration	Criticism		Central issue
3	Organization	Historical	Conceptual		Methodological
4	Perspective	Neutral representation		Espousal of position	
5	Audience	Specialized scholars	General scholars	Practitioners / politicians	General public
6	Coverage	Exhaustive	Exhaustive and selective	Representative	Central / pivotal

Fig. 3.1   Taxonomy of literature reviews following COOPER [104]

acquisition combined with the upcoming research area of ML providing new data analytics methods. Accordingly, this leads to the following research questions:

- RQ1: Which is the current state of the art in directly predicting lead times for manufacturing companies and which data and methods or algorithms are used?
- RQ2: How does the existing literature contribute to future research on direct lead time prediction?

3.4.2   Conceptualization of the Topic

Before conducting a review to synthesize knowledge from literature, according to the authors in [105] it is strongly recommended to acquire a priori knowledge about the topic, to identify potential areas where synthetized knowledge may be needed and to properly conduct the review. Based on the explanations and definitions provided in Section I and II and reviewing over 40 publications with an explorative approach we identified concepts most relevant to our field of observation and mapped them to the topic. So, it is ensured to use a wide range of key terms that are locatable within literature. As a result, we generated a concept map

[106] for lead time prediction (cf. Fig. 3.2). The concept map lists all relevant synonyms for the further literature search.

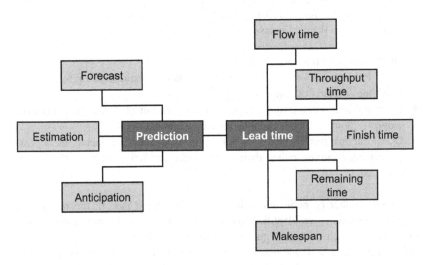

Fig. 3.2 Conceptualization map for lead time prediction according to the procedure of ROWLEY AND SLACK [106]

3.4.3   Literature Search

Based on the concept map the search terms were transferred into the following search string including Boolean operators and wildcards: ("predict*" OR "forecast*" OR "estimat*" OR "anticipat*") AND ("throughput time*" OR "flow time*" OR "remaining time*" OR "finish time*" OR "makespan*"). We used AND operators to exclude publications focusing on a single area of the search field only in order to increase the thematic relevance. The search strategy was enhanced by the elements of the STARLITE mnemonic framework [107]: We focus on journal articles and conference proceedings published in English between 1960 and 2019 in the electronic databases IEEE Xplore, Web of Science, EBSCO, ScienceDirect, and SpringerLink.

The application of the search string to the metadata title, abstract and key words, considering the additional criteria from the STARLITE mnemonic, identified a total of 18,697 publications in all databases. Afterwards, we followed the procedure given in the PRISMA flow diagram according to MOHER ET AL. [73] to consider relevant publications only. The procedure recommends to remove duplicates followed by a literature screening and detailed assessment of relevance based on the full text. The following quality criteria were defined for the screening and the detailed assessment:

- QC1: Addresses the domain of manufacturing.
- QC2: Publications are focusing on the prediction, estimation or forecast of lead times or parts of lead times.
- QC3: Publications focusing on algorithm development rather than methodological / domain specific applications are excluded.
- QC4: Publications focusing on job shop sequencing, queueing theory or performance curves rather than on a direct prediction of lead times are excluded.

The total number of publications included 3,786 duplicates. In the remaining 14,911 publications we identified various publications that do not comply with the applied search criteria. It turned out that some databases apply the search string to the full text in addition to title, abstract and key words. To comply with the search criteria, we additionally applied the search string to title, abstract and key words manually. After removing duplicates and the manual application of the search string a total number of 4,004 publications remain for the screening phase.

For screening the publications, we utilized a clustering approach by WEIßER AND SAßMANNSHAUSEN ET AL. [74] based on Natural Language Processing (NLP). Starting with a tokenization (word separation), the removal of stop words (stop words do not contain relevant information) and a TFIDF vectorization, a k-Means clustering is performed and the most relevant words (topwords) per cluster are identified. The topwords characterize each cluster and indicate its thematical relevance. We used title, abstract and key words without the search string as base for the clustering. Due to the resulting big text corpus we performed a dimensionality reduction by latent semantic analysis (LSA), as proposed by [108] and [109], to achieve better clustering results. Furthermore, to fully comply with the defined quality criteria, we did not solely rely on the topwords for excluding irrelevant clusters as proposed by WEIßER AND SAßMANNSHAUSEN ET AL. [74]. Based on the assumption of homogenous clusters, we have additionally taken a representative but random sample of publications of each cluster and read their full texts. Only if all of the publications in the sample do not match the quality criteria QC1-4, the whole cluster is assessed as irrelevant.

For the 4,004 remaining publications a clustering with ten clusters was performed and the topwords were extracted (cf. Table 3.1). The number of clusters was identified by applying the elbow method. Based on the analyzed samples and the topwords the clusters three, five and nine are assessed as relevant with a total number of 857 publications. Following the clustering, we analyzed the abstracts of all publications with respect to QC1-4. The remaining 367 publications were then further analyzed by reading the full text resulting in 39 relevant publications. With the relevant 39 publications we performed a forward and backward search, to identify models, theories and constructs that may not have been covered by the database search terms [110]. Thus, additional three relevant publications were identified, leading to the final data set of 42 publications for further analysis and synthesis in phase IV of the approach of VOM BROCKE ET AL. [72].

Table 3.1 Clusters with topwords, cluster size and assessed relevance

Cluster No.	Top Words	Cluster Size	Relevance
1	Model, data, based, system, using	1,349	Not relevant
2	Model, series, river, neural, network	463	Not relevant
3	Manufacturing, production, process, product	376	Relevant
4	Ensemble, precipitation, skill, model, weather	448	Not relevant
5	Abstract, copyright, may, users, abridged	193	Relevant
6	Flood, rainfall, model, river, warning	180	Not relevant
7	Traffic, series, network, model, term	68	Not relevant
8	Skill, enso, ocean, climate, sst	457	Not relevant
9	Inventory, demand, supply, chain, bullwhip	288	Relevant
10	Cancer, screening, patient, breast, survival	182	Not relevant

3.5   Section IV: Results

The intention of this theoretical overview is to bring relevant concepts into a superordinate structure, to map the contribution of literature to our problem statements, and to provide starting points for future research [103]. Therefore, publications with different concepts are analyzed and synthesized considering how they contribute to our research questions (cf. Sect. 3.4.1). Before performing the analysis and synthesis in Sect. 3.5.1 we define a framework as a base in Sect. 3.5.2.

3.5.1 Definition of the Framework

Setting up a framework is a common approach to structure literature as recommended by [111] and [112]. Our framework is separated in the following three dimensions (cf. Fig. 3.3):

3.5.1.1 Data Class

As a core differentiation we already mentioned the data class (cf. Section I and II). CRONJÄGER [113] divides the recorded data of manufacturing companies into order data, machine data, employee data and material data. *Order data* define all specific dates, times and quantities of individual orders. In our framework we will further include operation specific dates, times and quantities in the order data since an operation is part of an order. *Machine data* define all characteristics of

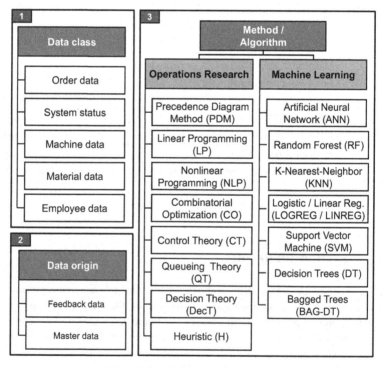

Fig. 3.3 Dimensions of developed framework

the machines that are used to process orders such as the machine ID, information about the tools or fault messages. *Employee data* contain information about the operators of the machines. This information is for example, the presence of employees or specific data such as the age or performance of an employee. *Material data* define all product characteristics of the product to be manufactured such as geometric specification, weights or the material itself. In addition, we identified publications that utilize information about the system status to directly predict lead times such as the stock level in intermediate storage or the capacity utilization of the machines (compare [17][114][115]). We have therefore added the *system status* as a fifth data class.

3.5.1.2 Data Origin

The analysis of the relevant publications showed that data used to directly predict lead times have various origins such as a planning data, a simulation or feedback data from a real production. For example, GOVIND AND ROEDER [116] generate input data for a direct prediction of lead times from a simulation. GRABENSTETTER AND USHER [117] consider historical data from a real production environment to directly predict lead times. Based on that we divided the second dimension of the framework *data origin* into the categories *feedback data* and *master data*. Feedback data describes data that was recorded in a real production environment during the production process. Master data are data used for planning without real feedback from a production environment. We included data that was generated from a simulation or whose origin is not further described within a publication in the category master data.

3.5.1.3 Method and Algorithm

Lead times can be predicted directly based on methods or algorithms from both research areas OR and ML (cf. section II). Since OR and ML are already established since many years, several overviews of these methods and algorithms are available in literature. For our framework we consider the basic works by ZIMMERMANN AND STACHE [118] and FEICHTINGER UND HARTL [119] to subdivide OR. They differentiate between Precedence Diagram Method (PDM), Linear Programming (LP), Nonlinear Programming (NLP), Combinatorial Optimization (CO), Control Theory (CT), Queuing Theory (QT), Decision Theory (DecT) and Heuristics (H). To subdivide ML we utilize the often-cited overview about supervised learning algorithms by CARUANA AND NICULESCU- MIZIL [120] to subdivide ML. They differentiate between Artificial Neural Networks (ANN), Logistic Regression (LOGREG), K-Nearest-Neighbor (KNN), Support Vector Machines (SVM), Random Forest (RF), Decision Trees (DT) and Bagged Trees

(BAG-DT). In addition to that we extended the field of Logistic Regression by Linear Regression (LINREG).

3.5.2 Analysis and Synthesis

Based on these defined dimensions we classified all publications accordingly and performed a descriptive analysis to identify the current state of the art in directly predicting lead times in manufacturing companies and in the used data classes and methods or algorithms (cf. RQ1). Additionally, we further deducted how the literature contributes to further research (cf. RQ2). A good overview of the development of a research area is given by the chronological development of the publications (cf. Fig. 3.4). Given the 42 identified publications, Fig. 3.4 shows an increasing number of publications focusing the direct prediction of lead times over time. Before the year 2000, we identified only three publications focusing the direct prediction of lead times, while the remaining 39 publications appeared after that date. Thus, a trend can be seen towards an increasing interest in the research area of directly predicting lead times.

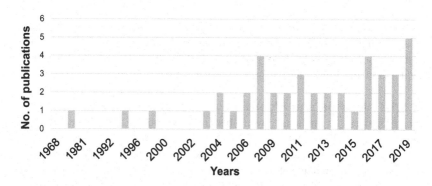

Fig. 3.4 Chronological development of publications

Next, we analyzed the dimensions of the framework (cf. Fig. 3.3) individually and subsequently combined two or more dimensions to identify common approaches and implications for further research. The following paragraphs are structured according to the considered dimensions.

3.5.2.1 Data Class

Looking at the data classes, it was noticeable that with a share of 95% of all publications, almost every author takes order data into account to directly predict lead times (cf. Fig. 3.5). JIA, ZHANG ET AL. [121], BERLEC AND GOVEKAR [122] or GRAMDI [123] for example use order data such as start and end dates of orders or order-specific processing times for the prediction of lead times. Therefore, order data are relevant for the direct prediction of lead times. Furthermore, the system status with a share of 62% of all publications is often used for direct predicting lead times. In contrast, machine and material data with a share of 21% and 5% respectively are used relatively rarely and employee data with a share of 0% have not been used for directly predicting lead times at all. One possible explanation for not using employee data could be, that due to data privacy restrictions employee data is not available for analysis. Furthermore, material data is commonly stored in the CAD-system, drawings or in the material master data in the ERP system, which might not be directly linked to the order data or system status. GYULAI, PFEIFFER ET AL. [124] and KARAGOLAN AND KARADEMIR [71] are the only authors who use material data such as dimensions or specifications of the product for directly predicting lead times. Machine data are used by WENG AND FUJIMURA [125], for example, in the form of the machine ID. LINGITZ, GALLINA ET AL. [70] use so-called 'equipment data' containing information about machines and tools to predict lead times without describing these data in more detail. The small proportion of machine, material and employee data suggests that either there is no or only a small relation between lead times and these data classes, or the connection has a low research interest in previous research. Since products in an engineer to order environment are designed individually and therefore the materials differ greatly in their characteristics, we see a high potential for further research considering material data as an input for directly predicting lead times.

Analyzing the number of used data classes in more detail reveals that 86% of all publications use two or less different data classes for directly predicting lead times (cf. Fig. 3.6). In case of using one data class only the majority of publications are considering order data like [126] or rarely system data like [114]. Machine and material data are not used solely. In case of using two or more data classes, order data is always included. With 40% the majority combines order data and the system status like [69]. Only a minority of 14% of all publications is using three data classes for directly predicting lead times combining order and system status with either machine data like [127] or with material data like [124]. Furthermore, it can be seen that in none of the publications more than three data classes are used. Since different combinations of three data classes have already

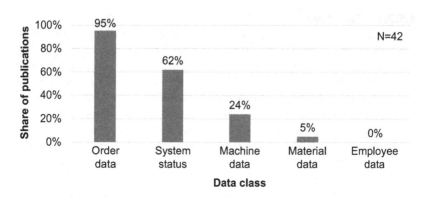

Fig. 3.5 Overview of used data classes

been successfully demonstrated, namely order data + system status + machine data and order data + system status + material data, it is also conceivable that a combination of all four data classes order data, system status, material data and machine data can provide good results in directly predicting lead times. Therefore, we see a high potential for further research in using three and more data classes for the direct prediction of lead times. Future researchers could, for example, develop a model using ML or OR in which, in addition to the system status and order data, they also use the material data to directly predict lead times.

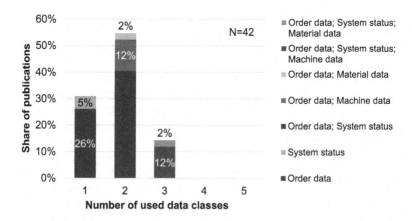

Fig. 3.6 Overview of quantity of used data classes

3.5.2.2 Method / Algorithm

Over time, the number of publications with ML increases continuously, whereas the number of publications with OR remains almost constant. In the case of ML 18 of the 23 publications were published after 2010. Therefore, the emerging trend of ML can also be seen in the research field of directly predicting lead times. In total the comparison of the research areas ML and OR shows with 55% only a slight majority in the area of ML compared to OR with 45% (cf. Fig. 3.7a). Looking at the ML methods and algorithms used in detail reveals that ANN (43% of all ML-publications), LINREG/LOGREG (30%), DT (26%) and RF (22%) were primarily used (cf. Fig. 3.7b). Furthermore, we identified authors using more than one approach within a publication to directly predict lead times. For example, ASADZADEH, AZADEH ET AL. [18] combine two approaches (ANN and LINREG) in one model, the authors in [128, 129] compare two approaches (ANN and DT) and the authors in [130] use a linear regressor (LINREG) to predict lead times. SCHUH, PROTE ET AL. [79] present a three-step procedure with a DT regressor for predicting order-specific interoperation times. GYULAI, PFEIFFER ET AL. [124] compare OR (e.g. Little's Law) and ML approaches and conclude that ML provides more precise results than OR. In their proposed model, a random forest approach is finally chosen because of a higher model accuracy for the available input data. Furthermore, a digital twin of the production environment is created to provide the ML model with quasi real production data for predicting lead times. Looking on the used OR methods and algorithms in detail reveals that Combinatorial Optimization (26% of all OR publications), Heuristics and Queuing Theory (both 21%) were primarily used (cf. Fig. 3.7c). For example, BERLEC AND STARBEK [16] use Combinatorial Optimization by setting up the lead times per operation of different orders in one vector per workstation and then randomly select and combine individual elements of the vectors to determine the total lead time of the order following a given processing sequence. In conclusion, in both research areas ML and OR specific methods and algorithms are used more frequently for directly predicting lead times while others like SVM or Control Theory are used rarely.

3.5.2.3 Data Class and Method / Algorithm

Combining the data class with the used method and algorithms reveals that order data is used in combination with all methods and algorithms (cf. Fig. 3.8). This deducts a general relevance of order data for directly predicting lead times, regardless of the method or algorithm used. The system status is used in 12 of 13 methods and algorithms for directly predicting lead times and can therefore be classified as generally relevant as well. Only decision trees are not used in

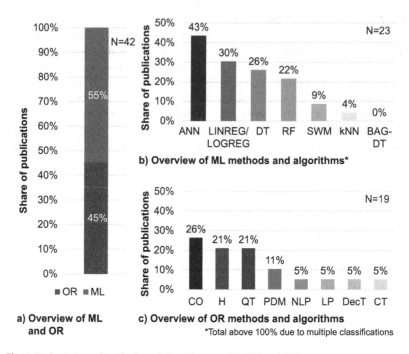

Fig. 3.7 Overview of methods and algorithms used in ML and OR

combination with the system status. Looking at the method of decision tree, we do not see any methodological reason for not using decision trees in combination with the system status. Considering machine data, it is noticeable that in more than 50% of cases combinatorial optimization (e.g., [131]) and ANN (e.g., [128]) are used. One possible explanation for this could be, that the information about several machines within the machine data need to be combined according to the corresponding processing sequence which is a typical application for combinatorial optimization and ANN. When using product data, it is noticeable again that only ANN in [71] and Random Forest in [124] are used to predict lead times. This either indicates that material data are not analyzable with other methods and algorithms, material data do not correlate with the directly predicted lead times or that material data has received less attention in prior research. Since there are already approaches with good results using material data for directly predicting lead times, we consider the second option, that material data do not correlate with lead times, as negligible.

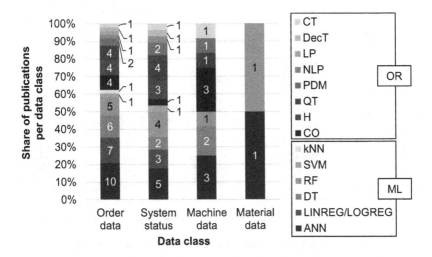

Fig. 3.8   Overview of data classes combined with used methods and algorithms

3.5.2.4   Data Origin and Method / Algorithm

Looking at the data origin only, we recognized an equal distribution of publications between feedback data and master data (cf. Fig. 3.9). Combining the used methods and algorithms with the data origin enables a more detailed view: Publications considering feedback data as base for directly predicting lead times utilize ML approaches with a share of 63% more frequently than OR. Here, most authors use ANN or LINREG/LOGREG. On the other side, OR approaches based on feedback data are dominated by CO. This leads to the insight that, from the field of ML, ANN and LINREG/LOGREG and, from the field of OR, CO are solid approaches for directly predicting lead times based on feedback data. KARAGOLAN AND KARADEMIR [71] for example perform a prediction of lead times using ANN and reach an accuracy up to 98.54% comparing the predicted lead times with the real lead times. In publications considering master data instead of feedback data with a share of 55% OR is used more frequently than ML. In detail ANN, RF, and QT are utilized almost equally. In conclusion, ML dominates the direct prediction of lead times based on feedback data whereas OR dominates the direct prediction of lead times based on master data. One possible explanation for this could be, that feedback data contain a larger amount of data sets which are predestined for ML, whereas the creation of master data is a manual and thus, expensive process which is suitable for OR.

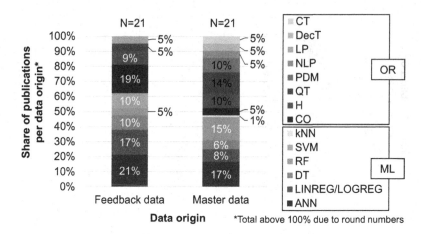

Fig. 3.9 Overview of data origin combined with used methods and algorithms

3.5.2.5 Data Class and Data Origin

Analyzing the combination of data class and data origin reveals a trend in the considered data origin depending on the used number of data classes (cf. Fig. 3.10). If only one data class is used for the direct prediction of lead times, almost 70% of the corresponding publications consider feedback data. If three data groups are used, the proportion of publications considering feedback data reduces to only 33%. This shows that the proportion of publications using feedback data decreases as the number of considered data groups increases. Since the number of data classes is an indicator for the model complexity, the identified trend implicates a decreasing use of feedback data for a direct prediction of the lead times with an increasing model complexity. Therefore, we see a high potential for further research focusing on higher model complexity with a larger number of data classes combined with feedback data.

The performed analysis and synthesis of the existing publications differentiated by the dimensions of our framework provided an extensive and detailed answer on RQ1. We identified data classes, data origins as well as methods and algorithms that are mainly used in the body of literature. We also identified implications for further research which we will summarize in the following section in detail.

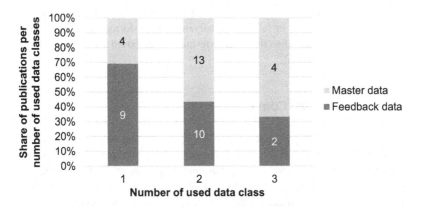

Fig. 3.10 Overview of data origin combined with the quantity of used data classes

3.6 Section V: Implications for Further Research

As already stated, all of the publications found in literature focusing the direct prediction of lead times could be classified with our developed framework (cf. section IV). By performing a descriptive analysis, we were able to identify common approaches that were used by the majority of researchers. Furthermore, we identified white spots and noticeable trends that indicate the need for further research (RQ2). Looking at the considered data classes we identified material data as an almost complete white spot in the research area of directly predicting lead times. Only few researchers present results in directly predicting lead times considering material data. With our review focus on the engineer to order production, where products often consist of a large number of components that are designed individually to achieve a tailor-made solution for the respective customer [1, 2], we see a high potential for further research considering material data in the direct prediction of lead times. Furthermore, we identified only few publications considering three or more data classes. Since disruptions in production systems are widely spread over various root causes [65], each of the different data classes might contain relevant information that correlate with the lead time. Additionally, we identified a decreasing number of publications using feedback data, if the number of used data classes increases. Feedback data contain the real information about the production system. Consequently, we see a high potential for further research considering three or more data classes for directly predicting lead times based on feedback data from a real production environment. Those few researchers focusing material data as input for directly predicting lead times only used ANN and RF so far. Thus, analyzing the performance of other methods and algorithms for directly predicting lead times based on material data is another research potential.

3.7   Section VI: Conclusion

In this article an SLR was conducted to determine the state of the art of directly predicting lead times with focus on engineer to order production. The lead time is one of the key factors for meeting customer requirements and predicting lead times can help to identify potential deviations from agreed delivery dates at an early production stage. Based on the identified deviations, the responsible person for production can then set counter measures to meet the due dates. The aim of this study was therefore to identify relevant data classes as well as methods and algorithms from the field of OR and ML used for directly predicting lead times within the body of literature. We conducted our research according to the SLR procedure model according to VOM BROCKE ET AL. [72] and integrated dedicated SLR concepts from other authors. Within the phase of literature search we identified a total of 18,697 publications, of which 42 publications were further considered in the core of our analysis. For the purpose of the selection of publications we utilized a clustering approach by WEIßER, SAßMANNSHAUSEN ET AL. [74] to allow a more efficient and target oriented scanning and filtering. In the subsequent analysis phase, a framework was developed to structure the considered publications followed by a descriptive analysis as the base to identify common approaches within the body of literature and to derive implications for further research.

A direct lead time prediction based on ML is a research field with increasing relevance. Concerning the considered data classes for the direct prediction, two data classes, namely order data and system status, are mainly used. Noticeable was the low usage of material data and feedback data in more complex models. From the field of ML, ANN and Regression models show high potential for further research in complex models considering material data and feedback data. With the performed detailed analysis all research questions stated in Section III were eventually answered.

We believe this study has both theoretical and practical implications. It provides academics with an overview of the state of the art of approaches for the direct prediction of lead times and indicates potential for further research. Furthermore, it can offer practical guidance to practitioners in selecting data classes as well as methods and algorithms to implement an approach for directly predicting lead times in their production environment.

3.8   Presentation of Own Contribution

All the work presented for the systematic research was carried out by me according to the procedure model of VOM BROCKE ET AL. This includes the definition of the review scope, the definition of the search string and the databases, the definition of quality criteria and research questions, the search of the literature corresponding to the search string, the evaluation of the literature according to the quality criteria, the development of the framework and the classification of the literature into this framework, the analysis of the literature by means of descriptive statistics as well as the subsequent synthesis and critical reflection of the literature. In addition, the transformation of the results of the systematic research into text and writing of the publication were entirely in my hands. All three co-authors Prof. Dr. Burggräf, Dr. Wagner and Mr. Koke contributed with ideas to the research concept and in the internal review process.

Publication II: Impact of Material Data in Assembly Delay Prediction—a Machine Learning-based Case Study in Machinery Industry

4

F. Steinberg, P. Burggaef, J. Wagner, and B. Heinbach, "Impact of material data in assembly delay prediction—a machine learning-based case study in machinery industry," *Int J Adv Manuf Technol*, vol. 120, 1–2, pp. 1333–1346, 2022, https://doi.org/10.1007/s00170-022-08767-3.

4.1 Abstract

Designing customized products for customer needs is a key characteristic of machine and plant manufacturers. Their manufacturing process typically consists of a design phase followed by planning and executing a production process of components required in the subsequent assembly. Production delays can lead to a delayed start of the assembly. Predicting potentially delayed components—we call those components assembly start delayers—in early phases of the manufacturing process can support an on-time assembly. In recent research, prediction models typically include information about the orders, workstations, and the status of the manufacturing system, but information about the design of the component is not used. Since the components of machine and plant manufacturers are designed specifically for the customer needs, we assumed that material data influence the quality of a model predicting assembly start delayers. To analyze our hypothesis, we followed the established CRISP-DM method to set up 12 prediction models at an exemplary chosen machine and plant manufacturer utilizing a binary classification approach. These 12 models differentiated in the utilization of material data—including or excluding material data—and in the utilized

F. Steinberg, *Machine Learning-based Prediction of Missing Parts for Assembly*, Findings from Production Management Research , https://doi.org/10.1007/978-3-658-45033-5_4

machine learning algorithm—six algorithms per data case. Evaluating the different models revealed a positive impact of the material data on the model quality. With the achieved results, our study validates the benefit of using material data in models predicting assembly start delayers. Thus, we identified that considering data sources, which are commonly not used in prediction models, such as material data, increases the model quality.

4.2   Section I: Introduction

Manufacturing companies are challenged to succeed in dynamic international markets requesting high-quality products, flexibility, on-time delivery, and a reasonable cost structure [64, 133, 134]. Here, short delivery times and adherence to delivery dates is a key factor to differentiate from competitors. A typical example of this is the machine and plant manufacturing industry producing complex products consisting of numerous components [1, 2]. Many of these components are customized enabling tailor-made solutions for the customers' requirements. In general, the manufacturing process of machine and plant manufacturers starts with the design of the product and components, followed by the production planning, the purchasing of raw materials, and the production process to manufacture the individual components needed in the subsequent assembly process. In parallel to the production process, the components required in the assembly are also purchased from suppliers. The task of the assembly is to assemble a product of higher complexity with predefined functions with a certain quantity of components in a partly multi-stage process in a given time [135]. Furthermore, in the assembly many material flows converge, leading to a high potential of delays [42]. Thus, an essential factor for meeting the delivery date is the start of the assembly on time and a prior timely supply of the components needed for assembly. Subsequently, components produced in the processes upstream of the assembly have a direct influence on the performance of the assembly process. Assuming that all components are required to start the assembly process, even a single component supplied behind schedule will lead to a delayed start of assembly [52]. To meet delivery dates, it would be helpful to predict these delayed individual components (we call them assembly start delayers) in the early stages of the manufacturing process. Based on an early prediction, measures such as close communication with the supplier, extra shifts to temporarily increase production capacity, or utilizing a different workstation can be derived to speed up the manufacturing process and thus, prevent assembly start delays.

With the increasing development of machine learning (ML) and the availability of big data, ML-based prediction models are becoming more and more established in the field of production planning and control. ML models have already been successfully applied to predict lead times of manufacturing processes [19] and to predict assembly start delayers [136]. Our previous research already showed that predicting assembly start delayers utilizing a binary classification is the recommended approach and outperforming approaches utilizing a lead time prediction to identify assembly start delayers [19]. Furthermore, when setting up and training a prediction model, the used data model has a central influence on the model quality of the prediction model [137, 138]. For example, BURGGRAEF ET AL. [19] have already discovered that material data defining all characteristics of the product to be manufactured such as geometric specification, weights, or the material itself are rarely used in ML-models to predict lead times.

Looking at the business process of a machine and plant manufacturer in contrast to the usage of material data in prediction models, it is noticeable, that the products of machine and plant manufacturers are typically tailor-made for each customer need [5, 139]. As the product's characteristics strongly influence the needed processes for its manufacturing [140], the design phase of machine and plant manufacturers including the material data specified within the design phase also has a non-negligible influence on the manufacturing process. Consequently, we assume that the usage of material data in a model predicting assembly start delayers has an impact on its model quality. Nevertheless, material data are currently only rarely used in prediction models. But, so far, a validation that the material data influence the respective model quality has not yet been performed.

Thus, our manuscript aims to set up an ML-based model for the prediction of assembly start delayers and to analyze and systematize the influence of material master data on the model quality. As a research method, we apply a case study at a machine and plant manufacturer. With the achieved results, our paper provides two main contributions:

- We developed a model to predict assembly start delayers utilizing a machine learning classification approach.
- We identified that material data influence the model quality of a model predicting assembly start delayers. However, there was only a slight influence.

Our paper is structured as follows. Section II first introduces the product structure and manufacturing processes in an engineer-to-order environment as well as available approaches to identify and predict assembly start delayers. Section III elaborates on our approach to quantify the impact of material data on the model quality predicting assembly start delayers utilizing ML. In Sect. IV, the results are presented and discussed. Section V critically reviews the limitations of our approach and the results obtained. Furthermore, the implications for further research are derived. Finally, a summary is given in the last section.

4.3   Section II: State of the art

The products of machine and plant manufacturers typically consist of several hundred to several thousand components. These are procured from suppliers or manufactured in the company's production facilities. Purchased components can be procured on an order-anonymous basis, such as for standard components, and an order-specific basis, such as for special and drawing components. The procurement of components from suppliers as well as the manufacturing of components in the in-house production belong to processes upstream of the assembly [32]. Since the assembly is a convergence point where several material flows converge, the risk of delays due to missing components is increased [46].

One established model to analyze converging material flows is the assembly flow element developed by SCHMIDT [9] with further developments and applications in the assembly flow diagram and supply diagram [9, 32]. In all models, the so-called completer is the last inflow to an assembly order and is therefore the component that was supplied last by the processes upstream of the assembly. A completer can be completed on time—before the planned start date of the assembly, or late—after the planned start of the assembly. A late finalization of a completer, therefore, leads to a delay in the start of assembly. In this manuscript, we define such components as "assembly start delayers" (see also section I). Assuming that all components are necessary to start the assembly, the schedule variance of the assembly start delayer determines the earliest possible start date of the assembly. Accordingly, a temporal acceleration of the manufacturing and/ or procurement process of an assembly start delayer has the biggest potential to push a delayed assembly start back to the target date. However, the supply diagram is primarily designed to analyze data relating to the past and to identify general issues such as an overall bad assembly supply situation in individual assembly areas. To derive case-specific countermeasures to accelerate individual production orders, further analysis is needed.

In production, typically scheduling techniques are used to derive order sequences and to calculate lead times of work orders used to determine the start dates and end dates of the respective orders and subsequently to determine the assembly start delayers [141]. The order sequence is defined according to certain rules considering for example the available production capacities, the technical requirements, the demand dates, and the system status [13, 14, 52]. Further, especially for remanufacturing systems, also environmental objectives are considered [142, 143]. To optimize the lead time of an order, the determination of its waiting time depending on the machine's utilization is essential [54]. Here, performance curves considering functional relationships between logistic parameters such as lead times, throughput, and stock play a key role [54, 55]. Nevertheless, deviations from the schedule may occur leading to an inaccurate determination of the assembly start delayers. Besides determining the assembly start delayers based on calculated lead times utilizing scheduling techniques, it is also possible to predict lead time directly. By predicting the lead times, completion dates can be determined early and deviations from the schedule can be detected [60]. In the past, many approaches for the prediction of lead times have been established. For example, CHENG AND GUPTA [68] investigated methods from the field of operations research (OR) such as Constant (CON), Random (RAN), or Total-Work (TWK). With the increasing development of ML, new methods for predicting lead times have emerged (see, for example, [69, 98, 99, 144]).

A systematic literature review conducted by BURGGRAEF ET AL. [19] has analyzed existing approaches focusing on the prediction of lead times in the research fields of ML and OR and classified them according to the three criteria data class, data origin, and used method/algorithm. Looking at the data class, the authors identified that the majority of publications examined use order data and information about the system status of the production system. In detail, 95% of their 42 publications examined use order data, and 62% use information about the system status. JAIN AND RAJ [121], BERLEC AND GOVEKAR [122] or GRAMDI [123] for example use order data such as start and end dates of orders or order-specific processing times for the prediction of lead times, whereas the authors in [69] and [114], for instance, use a combination of order data and information about the system status such as the machine utilization, processing times or the queue length. In contrast to the order data and information about the system, with 24% of the 42 publications examined, machine data are slightly less used. For example, the authors in [125] include the machine ID and the authors in [70] include the so-called 'equipment data' containing information about machines and tools in their prediction models. Further, BURGGRAEF ET AL. [19] identified GYULAI, PFEIFFER ET AL. [124] and KARAOGLAN AND KARADEMIR [71] with a portion of only 5% of the 42 publications examined as the only authors who include material data such as dimensions or specifications of the product in their prediction models. These findings highlight that material data were rarely used compared to order data, information about the system status, and machine data.

That being said, the business process of a machine and plant manufacturer typically hinges on tailor-made products for each customer need [5, 139]. Thus, the design phase in the business process and the associated documents herein can be said to have a non-negligible influence on the desired product. Furthermore, the product design is also the basis for the production planning determining the process to manufacture the respective components [140]. Accordingly, the material data specified in the design phase are also influencing the manufacturing process. Consequently, we assume that the usage of material data in a model predicting assembly start delayers has an impact on its model quality. Nevertheless, material data are currently only rarely used in prediction models.

Utilizing the findings of the systemic literature review in [19], the authors in [136] applied different ML algorithms on a total of 24 different prediction models on four different levels of detail to identify the modeling approach with the highest model quality in predicting assembly start delayers. Their models on the coarsest level of detail predicted assembly start delayers utilizing a binary classification. Their models on the three finer levels of detail predicted assembly start delayers via a prediction of different lead times (component lead times, order lead times, and operation lead times) utilizing a regression approach and subsequent postprocessing operations to identify the assembly start delayers. After training the 24 prediction models based on a real data set of a machine and plant manufacturer and evaluating their model quality, they identified the coarsest level of detail utilizing the binary classification as the best modeling approach. Thus, one of their findings was, that performing a binary classification to predict assembly start delayers outperformed the prediction of assembly start delayers based on a prior prediction of lead times utilizing a regression model. Accordingly, for our approach, applying a binary classification is recommended to predict assembly start delayers. Furthermore, the authors in [136] already used material data in all of their 24 prediction models leading to good results. Nevertheless, as they did not systematically analyze the impact of material on the model quality, there is still no analysis available proofing that material data have an impact on the quality of models predicting lead times.

In summary, there are models available for the prediction of lead times, but they are not explicitly used for the prediction of assembly start delayers. Currently, there is only one approach available focusing on the prediction of assembly start delayers in the field of machine and plant manufacturers comparing a direct prediction of assembly start delayers with an indirect prediction based on a previous lead-time prediction. But still, there is no analysis performed on the impact of material data on the quality of models predicting lead times.

Consequently, in this work, we will focus on investigating the influence of material data on the quality of models predicting assembly start delayers. This systemic analysis is completely novel compared to recent research. For this purpose, the following research question is posed, considering the previous explanations: "What effect does the use of material data have on the model quality of a model predicting assembly start delayers?" Following our argumentation that the products of machine and plant manufacturers are typically designed tailor-made to meet the specific customer needs and that the material data, therefore, characterize a product, we formulate the following working hypothesis: "The model quality for the prediction of assembly start delayers increases when utilizing material data."

4.4 Section III: Modelling approach

Examining an exemplary use case is an established approach in the field of machine learning, especially in lead-time prediction (see, for example, [70, 71, 126, 128]) and assembly start delayer prediction (see, for example, [136]). One motivation for examining an exemplary use case is to gain insights for real needs, such as the need of a manufacturing company, rather than to develop theories without practical relevance [145]. Accordingly, investigating an exemplary use case to answer our research question and to study our working hypothesis is an appropriate and established approach and thus, was our approach of choice. Furthermore, as this work extends our previous research in the prediction of assembly start delayers [136], we investigated the same case at the previously chosen representative machine and plant manufacturer.

The methodology used in this manuscript is following the established Cross Industry Standard Process for Data Mining (CRISP-DM) [146, 147] consisting of the six phases Business Understanding, Data Understanding, Data Preparation, Modeling, Evaluation, and Deployment.

4.4.1 Business understanding

In the Business Understanding phase, we derived objectives and requirements from a business perspective and converted them into a data mining problem. The objective from a business perspective was to prevent delays due to missing components in the final assembly so that the predefined due date of a customer order can be met. Early detection of components that have a higher tendency of late finishing in their preprocessing stages would be helpful to prevent a subsequent delay in the final assembly, as the production planer can accelerate the order in the preprocessing stages. The company under observation develops machines for steel production, which are made up of several hundred components. These components are both procured from suppliers and are manufactured in-house. An analysis carried out in the company beforehand showed that approx. 95% of the assembly start delayers are components produced in the company's production. Thus, the scope of our prediction model was constrained to the components manufactured in-house. In the process upstream of the assembly, these in-house components are processed by various machines for mechanical and welding operations.

In the prediction model, the components were classified as "assembly start delayer" (ASD) or "no-assembly start delayer" (NASD) which was identified as a suitable modeling approach in our previous research [136]. For this classification, a slightly modified version of the definition of the assembly start delayers given in chapt. 2 is applied: Instead of considering only one single assembly start delayer as a date determining factor for the assembly start according to the definition of BECK and SCHMIDT [9, 32] and thus, assigning the highest potential for improvement to this component, several assembly start delayers were considered for each assembly order. This extension is recommended, since considering only one assembly start delayer is not revealing whether this single one is an outlier or whether a large portion of the components is completed at a similar time. The modified assembly start delayer classification was defined as follows: If the schedule variance of a component is larger than or equal to 80% of the maximum schedule variance of all components of an assembly order, which is the schedule variance of the actual assembly start delayer, then this component is considered as an assembly start delayer. In detail, we utilized the formula

$$Class = \begin{cases} ASD;\ SV_{i,j} \geq 0,8 * SV_{j,max} \wedge SV_{i,j} > 0 \\ NASD;\ SV_{i,j} < 0,8 * SV_{j,max} \vee SV_{i,j} < 0 \end{cases} \tag{4.1}$$

to assign one of the two classes "assembly start delayer" (ASD) and "no assembly start delayer" (NASD) to every component i, where $SV_{i,j}$ is the schedule variance of component i of assembly order j, calculated by

$$SV_{i,j} = CD_{i,j} - TSD_j \tag{4.2}$$

where CDi,j is the calculated completion date of component i of assembly order j based on the predicted lead time of the prediction model and TSDj the target start date of assembly order j, and SVj,max the maximum schedule variance of all components of assembly order j, calculated by

$$SV_{j,max} = CD_{j,max} - TSD_j \qquad (4.3)$$

where CDj,max is the latest completion date of all components of assembly order j (the completion date of the respective completer).

The time of application of the prediction models (prediction time) and thus, the time of gaining knowledge about potential assembly start delayers should be as early as possible within the production process, as the production planer can accelerate the order in the manufacturing processes upstream of assembly stages. For the prediction models within this study, we set the date of order creation and thus, the completion of order planning as prediction time. At this point, all necessary information, such as bill of materials, operations, and machine assignments, are available.

Summarized, we converted the business objective to a binary classification problem. Subsequently, to answer the research question, and with our hypothesis that the model quality for the prediction of assembly start delayers increases when utilizing material data, we derived our data mining approach: We compared ML-based binary classification models using a data set including material data with ML-based binary classification models using the same data set but excluding the material data (cf. Fig. 4.1). For both cases, "including material data" and "excluding material data," we applied several ML-algorithms such as tree-based classifiers, support vector machines, or neural networks utilizing the Scikit-learn library or Keras library in Python (further details about the ML-algorithms used are explained in sect. 4.4.1). In total, 12 models were created, six per case utilizing different ML algorithms. Thus, with our approach we compared the performance of the different ML algorithms in both cases to identify the impact of material data on the model quality and the best performing ML algorithm by evaluating the achieved model qualities. Such a systemic analysis of the impact of material data on the model quality is completely new in recent research (see section II).

To evaluate the different achieved model qualities, we applied a confusion matrix, since the output of all ML models is the binary classification "assembly start delayer / no assembly start delayer." The evaluation of the model quality with a confusion matrix is an established method and has already been demonstrated in other studies (see, for example, [148, 149]). Based on the confusion

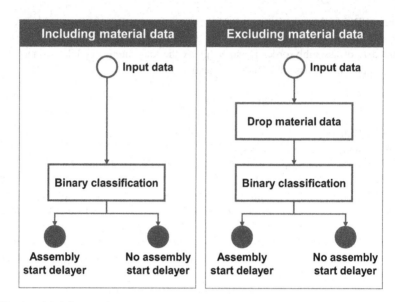

Fig. 4.1 Modeling architecture to quantify the impact of material data on the model quality

matrix, we calculated Matthew's correlation coefficient (MCC) and the F-score as established evaluation metrics to compare the performance of the different ML algorithms on both data sets. As recommended by the authors in [136, 150], the MCC considers the balanced ratios of all four confusion matrix categories and thus, is the most informative metric to evaluate a confusion matrix. Considering the MCC also ensured that our model was not just predicting the majority class in our data set, which is "no assembly start delayer." Furthermore, as recommended by the authors in [136] we considered the F-score as an evaluation metric since it is focusing on the prediction of positives (assembly start delayers) only, which is the most important category in our case of interest. For the F-score, we used the F_2-score in detail considering the recall two times as important as precision. This weighting is based on the assumption that it seems more important to identify as many of the actual assembly start delayers as possible, in case of doubt even more than exist, and to define acceleration measures for them, than not to identify individual assembly start delayers at all. By evaluating each ML model with these metrics, the impact of material data on the quality of a model predicting assembly start delayers can be determined. Furthermore, with the MCC and F_2-score, we use the same metrics as in our previous research [136] and thus ensure comparability.

4.4.2 Data understanding

In the data understanding phase, according to the authors in [147], we collected and analyzed the data to identify data quality problems and to develop a solid understanding of the dataset. The data were collected from the Enterprise Resource Planning (ERP) and Advanced Planning and Scheduling System (APS) of the plant and machine manufacturer under observation with a period under review of one year. In detail, we collected data from the four data classes order

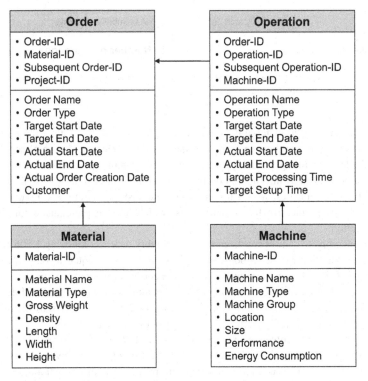

Fig. 4.2 Entity-relationship diagram with an excerpt of available features per data table

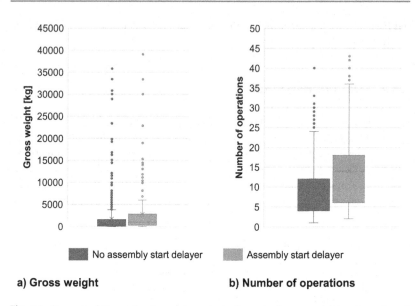

a) Gross weight   b) Number of operations

Fig. 4.3 Excerpt of the exploratory data analysis: Impact of gross weight and number of operations on assembly start delayers

data, machine data, material data, and system status, and thus follow the recommendation of the authors in [19]. The data export consisted of several separate CSV files containing assembly orders, the corresponding production orders and operation as well as information on the material and the systems status. To better join the different files, we set up an entity-relationship diagram (see Figure 4.2) enabling us to identify the primary keys, which are the prerequisite for their connection.

The complete dataset consisted of 356 assembly orders comprising 1,506 components supplied by the in-house production and thus, is equal to our previous research [136]. These 1,506 in-house components are manufactured by a total of 3,187 production orders comprising 15,772 operations. With our modified definition of an assembly start delayer, we had a total of 24% "assembly start delayers" and 76% "non-assembly start delayers" of all in-house components.

Further, as recommended by the authors in [147] we focused on gaining a better understanding of the data and developing first ideas of relevant data fields for the prediction of assembly start delayers by performing an exploratory data analysis. In detail, we utilized several graphical techniques such as boxplots, scatter plots, or Pareto charts. For example, we analyzed the distribution of the total number of operations needed to manufacture ASDs and NASDs (see Fig. 4.3a) showing a slight deviation between both classes. Components manufactured in more operations have a slightly higher tendency of becoming an ASD. As another example, we plotted the distribution of the gross weight of ASDs and NASDs as an initial study of the impact of material data (see Fig. 4.3b). ASDs have a slightly higher mean and median gross weight than NASDs. Heavier components may need extra handling effort and transport time and therefore have a higher tendency of becoming an ASD.

4.4.3   Data preparation

With the gathered understanding of the data, we continued with preparing the final dataset for training the models by transforming and cleaning the initial raw data. In detail, we continued to identify the relevant data field for the prediction models by performing a correlation analysis as recommended by the authors in [151]. Subsequently, after further data preprocessing operations such as discretization, decomposition, normalization, and aggregation (see, for details, [152, 153]), we defined the features for our data model resulting in 17 features, although not all features are applied in all models (see Table 4.1).

Since tree-based classifiers from the Scikit-learn library and neural networks from Keras library can only be trained on numerical variables in Python [154], the categorical variables such as "component name", "dispatcher" and "priority" were converted to Boolean values by performing One-Hot-Encoding. The number of features increases to a total of 375 features. Due to the One-Hot-Encoding, our data set was transformed into a sparse matrix containing equal information but in a higher dimensional room. This sparse matrix could for example hinder the optimization of a neural network, due to a not neglectable number of zeros as input of the model. Furthermore, the encoded features could have a dependency on each other. To investigate the correlations between the features, we created a 375 × 375 correlation matrix in form of a lower triangular leading to 71,631 individual correlation coefficients which were assigned to five bins of different correlation strengths (see Table 4.2) according to the established rules recommended by the authors in [155, 156]. Initially, 1.4% of all feature-pairs showed at least a moderate correlation a correlation coefficient higher than 0.5 and 1.5% of features pairs have low correlation. This indicates an existing dependency between our features. Thus, a Principal Component Analysis (PCA) was performed to avoid

Table 4.1   Features used in the prediction model

Data class	Feature	Including material data	Excluding material data
Order Data	Target lead time	X	X
	Total number of orders	X	X
	Total number of operations	X	X
	Target processing time	X	X
	Target setup time	X	X
	Order creation-delay	X	X
	Priority	X	X
	Operation type	X	X
	Dispatcher	X	X
	Number of production areas a component/order passes through	X	X
System status	Number of orders in system	X	X
Material data	Gross weight	X	
	Component name	X	
Machine data	Production area	X	X
	Workstation type	X	X
	Workstation number	X	X
	Workstation capacity	X	X

a sparse matrix and to reduce the dependencies between the features to ensure a good model quality. The improvement of the model quality by using a PCA has already been demonstrated in other studies (see, for example, [157]). By performing PCA, the 379 features were transformed into 46 principal components, which explain most of the variance of the original features. After performing PCA, we again performed a correlation analysis and assigned all correlation coefficients to the equal five bins (see Table 4.2), showing that all principal component pairs have a negligible correlation.

For training and evaluating the models, the dataset was divided into training and test sets with a ratio of 80% training data to 20% test data. In selecting the ratio, we followed established ratios. These are approx. 75%–80% training data to 25%–20% test data [158].

Table 4.2 Correlation between features before and after PCA following bin sizes of [155, 156]

Bin	Correleation coefficient	Before PCA	After PCA
Very high correlation	1.0 to 0.9 (–1.0 to –0.9)	0.3%	0.0%
High correlation	0.9 to 0.7 (–0.9 to –0.7)	0.6%	0.0%
Moderate correlation	0.7 to 0.5 (–0.7 to –0.5)	1.5%	0.0%
Low correlation	0.5 to 0.3 (–0.5 to –0.3)	6.5%	0.0%
Negligible correlation	0.3 to 0.0 (–0.3 to –0.0)	91.1%	100.0%

4.4.4   Modeling

The subsequent modeling phase covered the development of ML models and the calibration of the hyperparameters to optimal values [147]. All ML models predict assembly start delayers using a binary classification, which was identified as the best modeling approach in our previous research [136]. Thus, components are classified as "assembly start delayer" or as "no assembly start delayer." To ensure the comparability of all ML models, we chose the same set of ML algorithms on both data sets. In detail, we compared the performance of a Support Vector classifier (SVC), a Decision Tree (DT) classifier, a Random Forest (RF) classifier, an Adaptive Boosting (AdaBoost) classifier utilizing a DT-classifier as a base estimator, a Gradient Boosting (GB) classifier and a Multilayer Perceptron (MLP), since they are established approaches for binary classifications [159–161]. For the MLP, specifically, a double hidden layer feedforward net with stochastic gradient descent (SGD) optimizer was applied. The number of nodes was 46 nodes on the input layer to cover all input features after performing One-Hot-Encoding and PCA, 50 nodes on each hidden layer, and one node on the output layer for the binary classification. The number of hidden layers, the number of nodes on the hidden layers, and the activation function on the hidden layers were defined by continuous optimization of the model quality. In detail, we compared different network architectures ranging from one to ten hidden layers with 1 to 100 nodes per hidden layer. The best network structure was the above-mentioned double hidden layer net. As activation function for the output layer, a sigmoid function was chosen, which is particularly suitable for binary classifications [162]. For the hidden layers, we applied a ReLU function as activation function after comparing it with the sigmoid function, tanh function and He function regarding the reached model qualities. All classification models were implemented in Python

3.7 utilizing the Scikit-learn library and Keras library. An overview of the optimized hyperparameters used in each of the classification models is given in the appendix in Table 4.4 and Table 4.5.

In summary, we created 12 different prediction models to classify components as ASD or NASD. These 12 models differentiated in the utilization of material data—including or excluding material data—and in the utilized ML algorithm—six algorithms per material data case. The target was to quantify the effect of utilizing material data on the quality of a model predicting assembly start delayers while comparing different ML algorithms, which is a novel approach compared to recent literature. As metrics to evaluate the model quality, we used the MCC and F-Score based on a confusion matrix.

4.5   Section IV: Evaluation of model application

In the evaluation phase, the applied models were thoroughly evaluated to check whether they meet the targets of our data mining approach [147]: Quantifying the impact of material data on the quality of a model predicting assembly start delayers. Thus, we split the two data sets—including and excluding material data—into two separate train and test data sets. Subsequently, we trained and tuned all ML algorithms based on the train data sets and then evaluated the achieved model qualities based on the two test data sets. The results are documented in Table 4.3.

Upon evaluating the metrics, it is particularly noticeable that the models trained on the data set including material data achieved the best results. Furthermore, the best results per data set were both achieved by the GB classifier. With an MCC of 0.67 and an F_2-score of 77%, the GB classifier utilizing material data outperformed the GB classifier not utilizing material data with an MCC of 0.62 and an F_2-score of 71%. Thus, comparing the best ML model per data set already indicates a dependence of the model quality on the material data.

Additionally, we created boxplots showing the spread in the F_2-score and MCC of all ML models utilizing the two different data sets (see Fig. 4.4). With the boxplots, the overall dependency of the model quality on the material data independent of the considered ML algorithm was visualized. The distribution of the F_2-score and MCCs of the ML models trained on the dataset including material data differed from the respective distribution of the ML models trained on the dataset excluding material data. This indicated that, overall, the ML models trained on the dataset including material data performed better than those excluding material data. Thus, the comparison of the overall spread of the ML models

Table 4.3 Reached model qualities of all prediction models

Model	Including material data		Excluding material data		Deviation	
	MCC	F_2-Score	MCC	F_2-Score	MCC	F_2-Score
SVC	0.55	72%	0.47	67%	0.08	5%
DT	0.56	68%	0.52	66%	0.04	2%
RF	0.56	70%	0.50	68%	0.02	2%
GB	0.67	77%	0.62	71%	0.05	6%
AB	0.60	72%	0.56	68%	0.04	3%
MLP	0.59	73%	0.56	71%	0.02	2%

emphasizes the indication that material data have an impact on the quality of models predicting assembly start delayers.

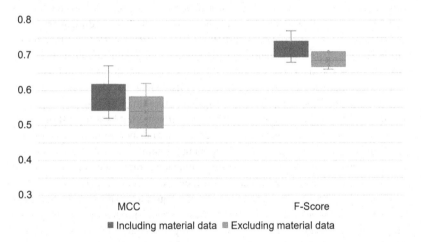

Fig. 4.4 Boxplot of MCC and F-Score for all prediction models on each of the four levels of details

Finally, we performed a statistical test to validate our working hypothesis. In detail, we performed two paired-samples t-tests, also referred to as dependent sample t-tests, both for MCC and F_2-score. This paired-samples t-test is used to assess whether the population means of two related samples differ. Thus, with the two paired-samples t-tests, we compared the means of the two samples 'ML models including material data' and 'ML-models excluding material data' individual for MCC and F_2-score. Additionally, we considered that the applied ML algorithms in each of the two samples were equal. Applying both tests revealed a p-value for MCC of approx. 0.003 and for the F_2-score of approx. 0.005. Consequently, since both p-values were less than 0.05, the difference between the two samples in both the MCC and F_2-score was statically significant. Accordingly, the impact of our considered material data on the model quality was statistically significant as well.

Consequently, the working hypothesis could be confirmed. The model quality significantly increased when material data were considered. However, in our case, there was only a slight increase in the MCC with an average of 0.04 and the F-score with an average of 3%. Thus, we further analyzed possible explanations for this small impact only and hypothesized prospects to further increase the benefit of utilizing material data. The reason for the small impact of material data observed could be that the considered material data—gross weight and component name—contain too little information to describe the characteristics of the components. Other information of the component such as dimensions, volume or number, and specification of features in the component's CAD model like drill holes, shaft shoulders, radii, or surface roughness could further increase the impact of material data. For example, the transportation, stocking, and handling effort of a component do not solely depend on its weight, but also other characteristics like dimensions and volume. For instance, the dimensions of a component determine whether the component can be easily transported by a forklift or crane, and thus, indicates an impact on an increase in transport times. Furthermore, the number and specification of a component's features indicate its complexity and need for special processing operation influencing the processing time. Thus, considering additional material data could increase the model quality.

In summary, we could answer our research question with our main contribution that the model quality of an ML-based model predicting assembly start delayers is significantly increasing when using material data. Thus, our study proved that models predicting assembly start delayers benefit from utilizing material data. In our exemplary case, we included the material data gross weight and component name in our prediction model significantly increasing the model quality. With these results, our approach is the first to systematically analyze the influence of material data on model quality in predicting assembly start delayers.

4.6 Section V: Limitations and implications for further research

In this work, we only considered one single machine and plant manufacturer as an exemplary case. Although a case-based approach is common in the field of machine learning (see, for example, [70, 71, 126, 128, 136]), the findings might remain case-specific and might not be generalizable. Accordingly, future research should validate the achieved findings considering additional machine and plant manufacturers in further case studies.

Nevertheless, in our work, we were able to show that material data have a positive influence on model quality for predicting assembly start delayers. However, the verifiable influence of the material data on the model quality was only small. We suspected the small range of data fields from the material data as a possible reason for this. Further material data could improve the model quality and thus strengthen the influence of the material data. Accordingly, future research should set up a model to predict assembly start delayers with additional material data.

The addition of further material data could also improve the generally low model quality. With a maximum MCC of 0.67 and a maximum F_2-score of 77%, the model quality is still too low for a successful practical application of the model, as there are still many false positive and false negative predictions. In general, the model quality depends on the input data, the utilized ML algorithm, and the complexity of the modeling approach [137, 163–165]. Together with our previous study [136], we already analyzed several different ML algorithms and modeling approaches. Thus, we infer that neither further optimization of the ML algorithm nor the modeling approach used is likely to lead to a significant improvement of the model quality. Instead, we infer that an enhancement of the input data could further improve the overall model quality, as the database also has an essential influence on the model quality [137, 138]. In our study, we already proved that material data influence the model quality. Consequently, we encourage further studies to consider additional data fields from the area of material data when setting up a model predicting assembly start delayers to further optimize the model.

Together with our previous work in the same research field [136], our findings observed are a good starting point in the prediction of assembly start delayers and the influence of material data on the model quality. As we could easily access the considered material data and integrate it into our data set, we added value to our model without much additional effort for data acquisition. Consequently, we have shown that it is worth also considering data, which might not have any influence on the model quality at first glance, and consequently is not commonly used. For future research in the field of applied machine learning, the elaboration of the database should be extended to other easily accessible data sources, even if they are not typically considered for the respective use case.

4.7   Section VI: Conclusion

At machine and plant manufacturers, the manufacturing process typically begins with the design of the product and its components before planning and executing the production process to manufacture the individual components needed in the subsequent assembly process. An essential factor for meeting a delivery date is the start of the assembly on time and a prior timely supply of the components needed for assembly. Subsequently, components produced in the processes upstream of the assembly have a direct influence on the performance of the assembly process. To meet delivery dates, we set up a supervised learning model to predict potentially delayed individual components (we call them assembly start delayers) in the early stages of the manufacturing process. Currently, machine learning models in the related area of lead time prediction typically include information about the system status, the machines, and the orders in their prediction model and do not consider material data [19]. As the design of a product is a central process for machine and plant manufacturers and the components are typically tailor-made to meet the customer's needs, we assumed that material data influence the model quality. Thus, we formulated the following working hypothesis: "The model quality for the prediction of assembly start delayers increases when utilizing material data." To verify the working hypothesis, we applied the established CRISP-DM procedure at an exemplary chosen machine and plant manufacturer. Here, we created 12 different prediction models to classify components as "assembly start delayer" or "no assembly start delayer." These 12 models differentiated in the utilization of material data—including or excluding material data—and in the utilized ML algorithm—six algorithms per material data case. The target was to quantify the effect of utilizing material data on the quality of a model predicting assembly start delayers while comparing different ML algorithms. As metrics to evaluate the model quality, we used the MCC and F-Score based on a confusion matrix.

Evaluating the different quality metrics of the 12 prediction models revealed a positive impact of the material data on the model quality. Thus, the working hypothesis could be confirmed. However, in our case, there was only a slight increase in the MCC and F-score. As a possible explanation for the small impact on the model quality, we suspect the limited information about the material considered in our model—gross weight and component's name only. Adding further information about the material such as dimensions, volume, or number and specification of features in the component's CAD model like drill holes, shaft shoulders, radii or surface roughness could further increase the impact of material data. Nevertheless, even with our limited consideration of material data. We verified, that utilizing data, which is commonly not used in prediction models increases the model quality.

In total, we successfully analyzed the impact of material data on the quality of models predicting assembly start delayers and gave insights into the performance of different modeling approaches. With our results, we achieved our two main contributions: First, we developed a model to predict assembly start delayers utilizing a machine learning classification approach. Second, we identified that material data influence the model quality of a model predicting assembly start delayers. However, there was only a slight influence. With our findings, for future machine learning approaches in the area of production planning and control, we recommend considering data sources apart from typically used data sources as well. We were able to show that even atypical data sources can contribute to an improvement of the model.

4.8   Appendix

The hyperparameters used in the prediction models were optimized utilizing a grid search and cross-validation algorithms (GridSearchCV) from scikit learn. Table 4.4 and Table 4.5 summarize the utilized hyperparameters in the different models on the four levels of detail.

Table 4.4 Hyperparameters of the prediction models including material data

Model			Hyperparameters			
	c	penalty	loss	max iter	dual	n estimators
SVC	1	l2	hinge	1000	True	-
	min samples split	min samples leaf	max features	max depth	learning rate	n estimators
DT	2	1	None	None	-	-
RF	2	1	70	8	-	800
GB	15	6	80	7	0.2	485
AB	30	15	80	1	0.2	1600
	momentum	dropout rate	batch size	epochs	learning rate	
MLP	0.9	0.2	64	450	0.001	

Table 4.5 Hyperparameters of the prediction models excluding material data

Model	Hyperparameters					
	c	penalty	loss	max iter	dual	n estimators
SVC	100	l2	hinge	1000	True	-
	min samples split	min samples leaf	max features	max depth	learning rate	n estimators
DT	2	3	None	None	-	-
RF	2	1	80	9	-	800
GB	16	6	80	7	0.2	694
AB	30	19	55	5	0.2	1600
	momentum	dropout rate	batch size	epochs	learning rate	
MLP	0.9	0.2	64	2000	0.001	

4.9   Presentation of own contribution

All the work presented within this publication was carried out by me. This includes the definition of research questions and hypotheses, the definition of the research approach including the set up to compare the performance of ML-models with different data sets—with or without material data—and different ML-algorithms, the collection, cleaning and aggregation of data from the manufacturing company under consideration, the feature engineering and setting up of the data model, the training and tuning of all ML-models including writing of the respective code in Python as well as the evaluation and critical reflection of the results. Further, transforming the work into text and writing of the publication were entirely in my hands. All three co-authors Prof. Dr. Burggräf, Dr. Wagner and Mr. Heinbach contributed with ideas to the research concept and in the internal review process.

Publication III: Machine Learning-based Prediction of Missing Components for Assembly—a Case Study at an Engineer-to-order Manufacturer

5

P. Burggräf, J. Wagner, B. Heinbach, and F. Steinberg, "Machine Learning-Based Prediction of Missing Components for Assembly—a Case Study at an Engineer-to-Order Manufacturer," *IEEE Access*, vol. 9, pp. 105926–105938, 2021, https://doi.org/10.1109/ACCESS.2021.3075620.

5.1   Abstract

For manufacturing companies, especially for machine and plant manufacturers, the assembly of products in time has an essential impact on meeting delivery dates. Often missing individual components lead to a delayed assembly start, hereinafter referred to as *assembly start delayers*. Identifying the assembly start delayers early in the production process can help to set countermeasures to meet the required delivery dates. In order to achieve this, we set up 24 prediction models on four different levels of detail utilizing different machine learning-algorithms—six prediction models on every level of detail—and applying a case-based research approach in order to identify the model with the highest model quality. The modeling approach on the four levels of detail is different. The models on the coarsest level of detail predict assembly start delayers utilizing a classification approach. The models on the three finer levels of detail predict assembly start delayers via a regression of different lead times and subsequent postprocessing operations to identify the assembly start delayers. After training of the 24 prediction models based on a real data set of a machine and plant

F. Steinberg, *Machine Learning-based Prediction of Missing Parts for Assembly*, Findings from Production Management Research ,
https://doi.org/10.1007/978-3-658-45033-5_5

manufacturer and evaluating their model quality, the classification model utilizing a Gradient Boosting classifier showed best results. Thus, performing a binary classification to identify assembly start delayers was the best modelling approach. With the achieved results, our study is a first approach to predict assembly start delayers and gives insights in the performance of different modeling approaches in the area of a production planning and control.

5.2   Section I: Introduction

Production companies are facing an ongoing change. They are challenged to assert themselves in international markets and to differentiate their products from other products available on the market in in terms of functionality, quality and price. Furthermore, the logistics performance, such as high adherence to delivery dates or short delivery and lead times, is becoming a key competitive factor [64, 133, 166]. A typical example for this are machine and plant manufacturers, whose products often consist of a large number of customized components to enable a tailor-made solution for the respective customer [1, 2]. To ensure high adherence to delivery dates and short lead times, the punctual assembly of a product is a central factor, as the product can only be delivered to the customer on time if it has been manufactured and assembled on time. The task of the assembly is to assemble a product of higher complexity with predefined functions with a certain quantity of components in a partly multi-stage process in a given time [135]. The manufacturing processes upstream of the assembly therefore have a direct influence on the performance of the assembly process, since a large number of material flows from different supply chains converge in the assembly process [42]. Often it is not possible to provide the required components on time and simultaneously. Under the assumption that all components required for assembly must be available at the start of assembly, the assembly process is subsequently delayed, if only one component is provided too late [52].

In order to avoid delays of the assembly start and thus to meet delivery dates, it would be helpful to predict potential missing components, we define those components as 'assembly start delayers', in early phases of the manufacturing process. By subsequently taking appropriate countermeasures, such as adding extra shifts in production or outsourcing of individual components, the assembly start delayers could be prevented. A central factor for the prediction of assembly start delayers is the lead time of the manufacturing processes upstream of the assembly. The aim of these manufacturing processes is the production of individual components. This is usually done in one or more sequentially executed orders, which in turn consist of one or more operations [167]. The lead time can therefore be considered at three different levels of detail: The component lead time, the order lead time and the operation lead time.

Due to the influence of the lead time on meeting the start of assembly, it seems obvious to predict assembly start delayers based on a lead time prediction. In addition, the lead time prediction can vary in the three levels of detail—component, order and operation lead time. It is also conceivable to predict the assembly start delayers directly via a classification, without a prior lead time prediction. This results in four different options with different level of detail to predict assembly start delayers. Thus, the aim of our paper is to set up a model for the prediction of assembly start delayers and to analyze and systematize the influence of the level of detail of the model on the model quality. As a research method we applied a case study at a machine and plant manufacturer. With the achieved results our paper provides two main contributions:

- We implemented machine learning models based on different algorithms to predict assembly start delayers.
- We identified the coarsest level of detail utilizing a binary classification as the best modeling approach.

Our paper is structured as follows. Section II first introduces the product structure and manufacturing processes in an engineer-to-order environment as well as available approaches for lead time prediction. Section III elaborates the prediction model to identify assembly start delayers utilizing different levels of detail. In section IV the results are presented and discussed. Section V critically reviews the limitations of the applied research method and the results obtained. Furthermore, the implications for further research are derived. Finally, a summary is given in the last section.

5.3 Section II: State of the art

The products of machine and plant manufacturers usually consist of several components. These are procured from suppliers or manufactured in the company's own production facilities [42]. Purchased components can be procured on an order-anonymous basis, such as for standard components, and on an order-specific basis, such as for special and drawing components. The procurement of components from suppliers as well as the manufacturing of components in the in-house production belong to processes upstream of the assembly [32]. Since the assembly is a convergence point where several material flows converge, the risk of delays due to missing components is increased.

One established model to analyze converging material flows is the assembly flow element developed by SCHMIDT [9] with further developments and applications in the assembly flow diagram and supply diagram [9, 32]. In all models, the so-called completer is the last inflow to an assembly order and is therefore the component that was supplied last by the processes upstream of the assembly. A completer can be completed on time—before the planned start date of the assembly, or late—after the planned start of the assembly. A late finalization of a completer therefore leads to a delay in the start of assembly. In this article we define such components as "assembly start delayer" (see also Chap. 1). Assuming that all components are necessary to start the assembly, the schedule variance of the assembly start delayer determines the earliest possible start date of the assembly. Accordingly, a temporal acceleration of the manufacturing and/or procurement process of an assembly start delayer has the biggest potential to push a delayed assembly start back to the target date. However, the supply diagram is primarily designed to analyze data relating to the past and to identify general issues such as an overall bad assembly supply situation in individual assembly areas. To derive case-specific countermeasures to accelerate individual production orders further analysis is needed.

The lead time of the processes upstream of the assembly has a central influence on meeting the target start date of the assembly and thus on meeting customer requirements. A single component is typically manufactured in one or more sequentially executed orders [167]. Consequently, we distinguish between a component lead time and an order lead time. Further, an order is typically subdivided into individual operations [11, 28]. Thus, we can differentiate between order lead times and operation lead times. The operation lead time is further subdivided into the operation time and interoperation time. As is well known, the interoperation time tends to have a higher share in the lead time than the operation time [28, 79].

In production, lead times are determined by setting up a production schedule taking into account the available production capacities, the technical requirements, the demand dates and the system status [13, 14, 52]. The order sequence is defined according to certain rules in order to calculate start and end dates of the orders at the workstations [53] and is one of the main applications for machine learning (ML) [97]. In addition to the calculation of the lead time based on scheduling, it is also possible to predict lead time directly. By predicting the lead times, completion dates can be determined early and deviations from the schedule can be detected [60]. In the past, many approaches for the prediction of lead times have been established. For example, CHENG AND GUPTA [68] investigated methods from the field of operations research (OR) such as Constant (CON), Random (RAN) or Total-Work (TWK). With the increasing development of ML, new methods for predicting lead times have emerged (see, for example, [69, 98, 99]).

A systematic literature review conducted by BURGGRÄF ET AL. [19] has analyzed existing approaches focusing on the prediction of lead times in the research fields of ML and OR and classified them according to the three criteria *data class*, *data origin* and used *method/algorithm*. Looking at the *data class*, the authors identified that the majority of publications examined use order data and information about the system status of the production system (see, for example, [122, 123]). In contrast, material data is rarely used, and employee data is never used to predict lead times. Given the fact that the products of machine and plant manufacturers are typically designed tailor-made to meet the specific customer needs, and that the material data therefore characterize a product, this information should be considered when predicting lead times. The authors in [71, 124] have already used material data utilizing artificial neural networks (ANN) and random forest (RF) for the prediction of lead times, but without using the primarily used information about the system status and machine data and furthermore not in the case of machine and plant manufacturers. According to BURGGRÄF ET AL. [19], there is a lack of prediction models for machine and plant manufacturers that use the primarily used data classes and material data for the prediction of lead times. ANN and RF have already proven successful in including material data in the prediction model. When looking at the *data origin*, the authors of [19] also identified that the use of real data strongly decreases with an increasing number of considered data classes. Thus, with increasing complexity of the prediction model they identified a lack of models using real data.

In addition to the selection of suitable data and a suitable approach, the level of detail of the model is crucial for a successful model application. According to the authors in [168], the level of detail refers to the system that the model represents (e.g. in the case of a model of a production line, the number of machines, components, etc. contained in the model), and not to the exact way in which the model is implemented (e.g. number of data fields used). Consequently, with respect to a model focusing on assembly start delayers, considering the lead time at the level of components, orders or operations would be possible levels of detail. An increase in the level of detail usually leads to a higher model accuracy, but with a degressive characteristic [165]. A 100% accurate model is only possible if the real system is fully mapped, which is typically not achieved [169]. Furthermore, an increase in the level of detail beyond a certain point can also lead to a less accurate model [165]. Therefore, in this work the impact of the degree of detail on the model quality will be investigated.

In summary, currently there is no model available for the prediction of assembly start delayers in the field of machine and plant manufacturers that considers the necessary complexity of the respective industry sector. There are models available for the prediction of lead time, but they are not explicitly used for the prediction of assembly start delayers. In addition, decisive data classes for machine and plant manufacturers such as material data are not used and there are also deficits in considering real data as the base for training the models. Furthermore, the level of detail of the model is not considered in any of the existing approaches.

In this work we will focus on investigating the influence of the level of detail of the modelling on the model quality. For this purpose, the following research question is posed, considering the previous explanations: "How does the level of detail of the modelling affect the model quality to predict assembly start delayers?" Considering the argumentation of the authors in [165], we formulate the following working hypothesis: "The model quality for the prediction of assembly start delayers increases with a finer level of detail."

5.4   Section III: Modelling approach

A case-based research approach is used to answer the research question and to investigate the working hypothesis. A case-based research approach is an objective, detailed investigation of a current phenomenon where the researcher has little control over real events [170]. One motivation for a case-based research approach is to gain insights for real needs, for example the needs of manufacturing companies, rather than to develop theories without practical relevance [145]. Furthermore, a case-based research approach has already been successfully applied in the area of lead time prediction (see, for example, [70, 71, 126, 128]. Although the research question focuses on the prediction of assembly start delayers and not on a lead time prediction, the lead time is one of the central factors for an assembly in time and thus a related research area. Accordingly, a case-based research approach is an appropriate method to answer the research question and to investigate the working hypothesis.

As representative case for the case-based research approach, a machine and plant manufacturer was chosen. A product of this company usually consists of several hundred components and is then used in steel production. An analysis carried out in the company beforehand showed that approx. 95% of the assembly start delayers are components produced in the company's own production. Accordingly, components procured from suppliers were not considered in the developed prediction model. Thus, the scope of this article is limited to the production of components in in-house manufacturing.

5.4.1 The prediction model

To answer the research question, 24 ML-models were created in total, which differ in their level of detail (see Fig. 5.1) and the utilized ML-algorithm. The models at the different levels of detail are independent of each other, but all pursue the same goal: the prediction of assembly start delays. To achieve this goal, each model comprises various operations. Further, we compared the performance of the different ML-algorithms on each level of detail to identify the best performing ML-algorithm by evaluating the achieved model qualities.

The first and coarsest level of detail (1) is the prediction of assembly start delayers using a binary classification. On this level of detail, components are classified directly as "assembly start delayer" or as "no assembly start delayer". On the levels of detail (2)-(4) the assembly start delayers are indirectly predicted based on a lead time prediction. With increasing level of detail, a finer granular consideration of the lead time, according to the definition of lead times by the authors in [28] (see Chap. 2), is used for the prediction. Consequently, the component lead time is used on the second level of detail (2), the order lead time at the third level of detail (3) and the operation lead time at the fourth and thus finest level of detail (4).

The detailed explanation of the operation principals including the ML-algorithms used on the four levels of detail (see Table 5.1) is first given for the coarsest level of detail (1). Afterwards the operation principal of the levels of detail (2)-(4) is explained. In the explanation the levels of detail (2)-(4) are considered together since their operation principal and the ML-algorithms used are analogous and differs only in the considered lead time. The prediction models on all levels of detail were implemented in Python 3.7 utilizing the scikit-learn library.

In the models on the coarsest level of detail (1) (see Fig. 5.2), we compared the performance of a Support Vector classifier (SVC), a Decision Tree (DT) classifier, a Random Forest (RF) classifier, an Adaptive Boosting (AdaBoost) classifier utilizing a DT-classifier as base estimator, a Gradient Boosting (GB) classifier and an

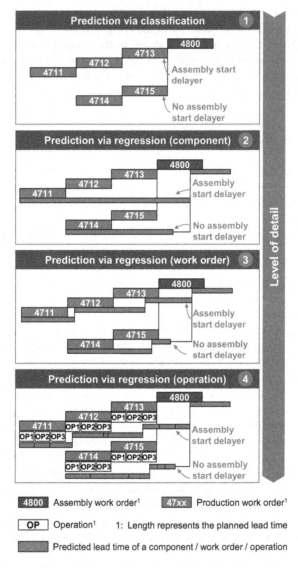

Fig. 5.1 Concept of the prediction models on four levels of details

Table 5.1 ML-algorithms utilized in the considered four levels of detail

ML-algorithm	Level of detail 1 Classification	Levels of detail 2–4 Regression
Support Vector Machine (SVM)	X	
Linear Regression (LR)		X
Decision Tree (DT)	X	X
Random Forest (RF)	X	X
Adaptive Boosting (AdaBoost)	X	X
Gradient Boosting (GB)	X	X
Artificial Neural Network (ANN)	X	X

ANN, since they are established approaches for binary classifications [159–161]. For the ANN, specifically, a single hidden layer feedforward net with a sigmoid function as activation function and a stochastic gradient descent (SGD) optimizer was applied. The sigmoid function as activation function is particularly suitable for binary classifications [162]. The number of nodes was 46 nodes on the input layer to cover all input features after performing One-Hot-Encoding, 16 nodes on the hidden layer and two nodes on the output layer to ensure the binary classification 'assembly start delayer' and 'no assembly start delayer'. The number of hidden layers, the number of nodes on the hidden layers and the activation function on the hidden layers were defined by a continuous optimization of the model quality. In detail, we compared different network architectures ranging from one to ten hidden layers with 1 to 100 nodes per hidden layer and different activation functions on the hidden layers such as ReLu function, sigmoid function, tanh function and He function. The best network structure was the above mentioned single hidden layer net. An overview of the optimized hyperparameters used in each of the classification models is given in the appendix in Table 5.5.

For the classification a slightly modified version of the definition of the assembly start delayers given in section II is applied: Instead of considering only one single assembly start delayer as a date determining factor for the assembly start according to the definition of BECK AND SCHMIDT [9, 32] and thus assigning the highest potential for improvement to this component, several assembly start delayers were considered for each assembly order. We recommend this extension, since considering only one assembly start delayer is not revealing whether this single one is an outlier or whether a large portion of the components are completed at a similar time. The modified assembly start delayer classification was defined as follows: If the schedule variance of a component is larger or equal to 80% of the maximum schedule variance of all components of an assembly order, which is the schedule variance of the actual assembly start delayer, then this component is considered as an assembly start delayer.

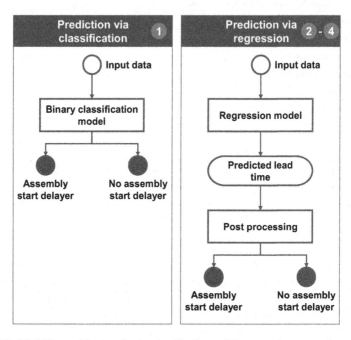

Fig. 5.2 Modelling architecture for the classification and the regression approaches

The models on the levels of detail (2)-(4) (see. Fig. 5.2) are based on a lead time prediction using a regression approach. Here we compared the performance of a linear regression (LR), a DT-regressor, a RF-regressor, an AdaBoost-regressor utilizing an DT-regressor as base estimator, a GB-regressor and an ANN, since they are established approaches for regression which have already been successfully applied in lead times prediction [19]. For the ANN, specifically, a single-hidden-layer feedforward net with a rectified linear unit as activation function utilizing a Keras regressor was applied. This activation function is particularly suitable for the prediction of lead times, since its output is limited to positive values only (negative lead times are not plausible), and it is an established activation function for regression models in ML [171]. The number of nodes was 46 nodes on the input layer on the level of detail (2) and 45 nodes on the levels of detail (3) and (4) to cover all input features after performing One-Hot-Encoding, 12 nodes on the hidden layer and one node on the output layer to enable the lead time as output of the regression model. The network architecture was also continuously optimized by comparing different numbers of hidden layers, number of hidden nodes on the hidden layers and activation functions on the hidden layers. In detail, we followed the same procedure as for the classification models and compared network architectures ranging from one to ten hidden layers with 1 to 100 nodes per hidden layer and different activation functions on the hidden layers such as ReLu function, sigmoid function, tanh function and He function. The best network structure was the above mentioned single hidden layer net. An overview of the optimized hyperparameters used in each of the regression models is also given in the appendix in Table 5.6—Table 5.8.

However, based on the predicted lead time only, it is not yet possible to make a statement about a potential assembly start delayer. In order to be able to identify the assembly start delayers at the levels of detail (2)-(4), additional subsequent operations were implemented (cf. "postprocessing" in Fig. 5.2): A completion date was calculated individually for each component, starting from a fictitious start date and using their respective predicted lead times. The fictitious start date was assumed to be either the target start date of the component or, if the target start date was already in the past at the time of creation of the corresponding production order and thus could not be realized, the date of the order creation and thus the completion of order planning. Typical examples of components for which the target start date at the time of order creation is in the past are supplement orders. At the levels of detail (3) and (4), an intermediate step was performed before calculating the completion date based on a fictitious start date: All predicted lead times (order lead times or operation lead times) of the respective component were summed up to a component lead time. Subsequently, at all three finer levels of detail (2)-(4), the assembly start delayers were determined according to the modified assembly start delayer logic based on the prior calculated completion dates of all components of an assembly order. In detail we utilized the formula

$$Class = \begin{cases} ASD; SV_{i,j} \geq 0, 8*SV_{j,max} \wedge SV_{i,j} > 0 \\ NASD; SV_{i,j} < 0, 8*SV_{j,max} \vee SV_{i,j} < 0 \end{cases} \tag{5.1}$$

to assign one of the two classes "assembly start delayer" (ASD) and "no assembly start delayer" (NASD) to every component i, where $SV_{i,j}$ is the schedule variance of component i of assembly order j, calculated by

$$SV_{i,j} = CD_{i,j} - TSD_j \tag{5.2}$$

where $CD_{i,j}$ is the calculated completion date of component i of assembly order j based on the predicted lead time of the prediction model and TSD_j the target start date of assembly order j, and $SV_{j,max}$ the maximum schedule variance of all components of assembly order j, calculated by

$$SV_{j,max} = CD_{j,max} - TSD_j \tag{5.3}$$

where $CD_{j,max}$ is the latest completion date of all components of assembly order j.

After performing the subsequent operations, the output of the models on the three finer levels of detail (2)-(4) is also "assembly start delayer" or "no assembly start delayer".

The applied procedure in the regression models, first to predict a lead time and, based on this, to calculate the completion dates of the components based on a fictitious start date, seems to be a cumbersome process. One could also think of directly predicting the completion dates without the workaround of predicting lead times. However, a direct prediction of the completion dates of the components is not possible with a supervised learning approach: Supervised learning is based on historical training data. If completion dates were directly predicted based on this historical training data, all completion dates would be in the past and not in the future. Therefore, predicting lead time is used as a workaround, as lead times depend on technical and organizational factors such as the available capacity or the required processing order, whereas they are usually independent of the considered date.

5.4.2 The data model

The data model on all four levels of detail consists of the four data classes *order data, machine data, material data* and *system status* and thus follows the recommendation of the authors in [19]. For the data acquisition, we followed the procedure of FAYYAD ET AL. [172] and HAN ET. AL. [152] consisting of the four steps *selection, pre-processing, reduction* and *transformation*. In the *selection*, the data for predicting the assembly start delayers was selected from the Enterprise Resource Planning System (ERP) and Advanced Planning and Scheduling System (APS) of the machine and plant manufacturer under consideration. According to the recommendation of the authors in [173] we included experts from the

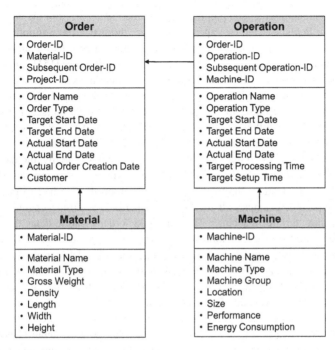

Fig. 5.3 Entity-relationship diagram with an excerpt of available features per data table

machine and plant manufacturer in this process. We considered all data fields that, based on the experience of the experts, have an impact on production orders meeting the target start date of the assembly and thus should be included in the prediction model. In addition, further data fields were selected which the experts classified as only potentially relevant. The data export included assembly orders, the corresponding production orders and operation as well as information on the material and the systems status. The period under review was set to one year. In the *preprocessing* which followed the selection, the data set was corrected by formatting individual data fields and cleaning up data gaps. Here, we also included the company´s experts to avoid deleting data e.g. with data gaps. In the preprocessing we also analyzed the data structure and combined the different raw data tables, which were basically separate csv-files, to one data model. For this, we set up an entity-relationship diagram (see Fig. 5.3) enabling us to identify the primary keys, which are the prerequisite for the connection.

After completing the data preprocessing, in the *reduction*, from the large number of data fields selected by expert knowledge, those that have an influence on the start of assembly were selected. For this purpose, a correlation analysis was performed according to the recommendation of the authors in [151]. In the final step, the *transformation*, the data fields were modified in order to define suitable features for the four prediction models. Here we applied typical methods such as discretization, decomposition, normalization, and aggregation (see, for details, [152, 153]). In the following, the transformation of the data fields 'workstation type' from the data class *machine data* and the data field 'order creation-delay' from the data class *order data* are explained as examples. Initially, the data field 'workstation type' was a free text field with many different characteristics. For the definition of the feature, the workstations were grouped according to their processing type. For example, all machines that perform a turning operation were grouped into 'turning machine'. The data field 'order creation-delay' has been calculated based on the deviation between the target start date of an order and the actual date of the order creation and thus indicates a delay in the order creation. In total, the application of the methodology of FAYYAD ET AL. [172] and HAN ET. AL. [152] results in 17 features, although not all features are applied at all levels of detail (see Table 5.2).

After performing One-Hot-Encoding for each level of detail we increased the number of features to a total of 375 features on the levels of detail (1) and (2) and 374 features on the levels of detail (3) and (4) due to many values in the categorical features. We further evaluated the dependence between the features by creating a 375×375 correlation matrix in form of a lower triangular matrix for the coarsest level of detail leading to 71.631 individual correlation coefficients. To get an overview of the overall correlation in our dataset we assigned all correlation coefficients to bins of different correlation strengths following the established rules for interpreting correlation coefficients [155, 156] leading to a total of five bins. Finally, we calculated the share of the individual bins in the number of all correlation coefficients (cf. Table 5.3). Based on the overview, we identified that 1.4 % of all correlation coefficient show at least a moderate correlation. This indicates an existing dependency between our features. Thus, a Principal Component Analysis (PCA) was performed to reduce the dependencies between the features and to ensure a good model quality. The improvement of the model quality by using a PCA has already been demonstrated in other studies (see, for example, [157]). By applying a PCA, we identified 46 principal components (PC) on the levels of detail (1) and (2) and 47 PC on the levels of detail (3) and (4) as an appropriate number of PC. After performing the PCA, we again performed a correlation analysis and assigned all correlation coefficients to the equal five bins (cf. Table 5.3) showing that the PCA eliminates the dependency between the features.

Table 5.2 Features used in the prediction model

Data class	Feature	Level of detail			
		1	2	3	4
Order Data	Target lead time	X	X	X	X
	Total number of orders	X	X		
	Total number of operations	X	X	X	
	Target processing time				X
	Target setup time				X
	Order creation-delay	X	X	X	X
	Priority	X	X	X	X
	Operation type	X	X	X	X
	Dispatcher	X	X	X	X
	Number of production areas a component/order passes through	X	X	X	
System status	Number of orders in system	X	X	X	X
Material data	Gross weight	X	X	X	X
	Component name	X	X	X	X
Machine data	Production area	X	X	X	X
	Workstation type	X	X	X	X
	Workstation number	X	X	X	X
	Workstation capacity	X	X	X	X

Our final dataset consisted of 356 assembly orders comprising 1,506 components supplied by the in-house production. Of course, the in-house-components were only a subset of all components needed for assembly. Components purchased from suppliers were excluded based on an analysis previously performed by the machine and plant manufacturer showing that the in-house-components are predominantly responsible for a delayed start of the assembly. These 1,506 in-house-components are manufactured by a total of 3,187 production orders comprising 15,772 operations. With our modified definition of an assembly start delayer we had a total of 24 % "assembly start delayers" and 76 % "non-assembly start delayers" of all in-house-components.

Table 5.3 Correlation between features before and after PCA following bin sizes of [155, 156]

Bin	Correleation coefficient	Before PCA	After PCA
Very high correlation	1.0 to 0.9 (−1.0 to −0.9)	0.3 %	0.0 %
High correlation	0.9 to 0.7 (−0.9 to −0.7)	0.6 %	0.0 %
Moderate correlation	0.7 to 0.5 (−0.7 to −0.5)	1.5%	0.0 %
Low correlation	0.5 to 0.3 (−0.5 to −0.3)	6.5%	0.0 %
Negligible correlation	0.3 to 0.0 (−0.3 to −0.0)	91.1%	100.0%

5.4.3   Training/test split and prediction time

After defining the data model and before training the prediction models, the data set was divided into training and test data with a ratio of 80% training to 20% test data. In selecting the ratio, we followed established ratios. These are approx. 75%-80% training data to 25%-20% test data [158]. When splitting the data, we ensured that the components of one assembly order are not separated. Thus, the data-subsets (training and testing) always contain the complete bill of materials of an assembly order produced in in-house production including all corresponding production orders and operations. By this, we ensured that the prediction model is subsequently able to predict the actual assembly start delayers.

The time of application of the prediction models (prediction time) and thus the time of gaining knowledge about potential assembly start delayers should be as early as possible within the production process, so that companies have as much time as possible to implement acceleration measures. For the four models within this study, we set the date of order creation and thus the completion of order planning as prediction time. At this point, all necessary information, such as bill of materials, operations and machine assignments are available.

5.4.4   Evaluation of the model quality

To evaluate the model quality of all models we applied a confusion matrix, since the output on all four levels of detail is the binary output "assembly start delay-er" or "no assembly start delayer". The evaluation of the model quality with a confusion matrix is an established method and has already been demonstrated in other studies (see, for example, [148, 149]). Following the authors in [150] we used the Matthew's correlation coefficient (MCC) [174] as an evaluation metric, since it considers the balance ratios of all four confusion matrix categories and thus is the most informative metric to evaluate a confusion matrix. Considering the MCC also ensured that our model was not just predicting the majority class in our data set, which is "no assembly start delayer". Furthermore, we considered the F-score, precision and recall [149] as evaluation metrics, since they focus on the prediction of positives (assembly start delayer) only, which is the most important category in our case of interest. In the F-score we weighted the recall twice as high than the precision, deviating from a regular harmonic mean. This weighting is based on the assumption that it seems more important to identify as many of the actual assembly start delayers as possible, in case of doubt even more than exist, and to define acceleration measures for them, than not to identify individual assembly start delayers at all. By evaluating each prediction of the four different levels of detail using these metrics, the dependence of the model quality on the level of detail of the modeling can be determined.

Besides considering the metrics MCC and F-score only, one could think to consider the model accuracy, which is the portion of correctly predicted assembly start delayers and non-assembly start delayers to all predictions, as well. Nevertheless, the model accuracy is not a suitable metric for our study, as there is an imbalance between assembly start delayers and non-assembly start delayers (in our dataset 24 % to 76 %). This is due to the definition of assembly start delayers, according to which the assembly start delayers are only a small portion of all components of an assembly order. A typical example would be an assembly order consisting of 100 components, 5 of which are assembly start delayers. If the model would predict "non-assembly start delayers" for all components, the accuracy would be 95 %. Nevertheless, none of the assembly start delayers, thus none of the critical components, would have been identified and consequently the goal of the prediction model would not have been reached. With the original definition of the completer given by the authors in [9, 32], according to which there is only one assembly start delay per assembly order, this imbalance would have been even stronger. Therefore, we only considered the MCC and F-score as suitable metrics to evaluate the model quality for the prediction of assembly start delayers.

In summary, we implemented and compared 24 prediction models on four different levels of detail (six models per level). The target was first, to identify the ML-algorithm reaching the highest model quality per level of detail and based on that, to identify the dependence of the model quality on the level of detail of the modeling. The models on the coarsest level of detail (1) utilizing a classification to directly predict assembly start delayers differ strongly from the models on the finer levels of detail (2)-(4) utilizing a lead time prediction based on a regression to predict assembly start delayers. The models on the three finer levels of detail (2)-(4) only differ in the utilized ML-algorithms and the considered lead time, which becomes increasingly finer with the level of detail: from the component lead time to the order lead time to the operation lead time. In all 24 models the output was the binary classification "assembly start delayer" or "no assembly start delayer". To enable the binary classification for the regression models on the levels of detail (2)-(4) the model output was postprocessed. As metrics to evaluate the model quality we used the MCC, F-Score, precision and recall based on a confusion matrix.

5.5   Section IV: Results

After the definition of the concept, the data model and the model evaluation, we trained the prediction models on our data set. Hyperparameter tuning was performed to optimize the model quality in the best possible way (cf. Table 5.5— Table 5.8). Subsequently, the confusion matrices were created for each model on the different levels of detail to determine the model quality. Based on the respective confusion matrix, the metrics MCC, F-score, precision and recall were calculated for each model (cf. Table 5.4). These metrics enabled us to determine the best performing ML-algorithm on each level of detail and the dependence of the model quality on the different level of detail.

Evaluating the metrics on the various levels of detail, it is particularly noticeable that the best result was achieved at the coarsest level of detail (1): The direct prediction of assembly start delayers utilizing a GB-classifier achieves the highest model quality with an MCC of 0.65 and an F-score of 75 %. With MCCs of approx. 0.3 to 0.4 and F-scores of approx. 50 % to 60 %, the best models on the finest three levels of detail (2)-(4) do not reach the result of the best model on the coarsest level of detail (1). Considering the levels of detail (2)-(4) the MCC, F-Score, precision and recall of the best performing model on each level increases with a finer level of detail. Thus, the model quality of the best regression models increases with a finer level of detail but still lower than the model quality of the best classification model, which was on the coarsest level of detail (1).

Furthermore, we created boxplots for the four levels of detail to visualize the spread of all models in the MCC and F-Score within the respective levels of details and the dependence of the model quality on the level of detail (cf. Fig. 5.4). It is particularly noticeable, that the models on the level of detail (1) strongly differ from the models on the levels of detail (2)—(4) emphasizing that the classification approach outperforms the regression approaches. In addition, the

Table 5.4 Reached model qualitites of all prediction models

Level of detail 1					Level of detail 2				
Model	MCC	F-Score	Precision	Recall	Model	MCC	F-Score	Precision	Recall
SVM	0.55	72 %	56 %	78 %	LR	0.22	38 %	33 %	40 %
DT	0.56	68 %	60 %	71 %	DT	0.30	43 %	40 %	44 %
RF	0.52	60 %	64 %	59 %	RF	0.10	26 %	24 %	27 %
GB	**0.65**	**75 %**	**67 %**	**75 %**	GB	0.22	35 %	35 %	35 %
AdaBoost	0.58	71 %	61 %	74 %	AdaBoost	0.13	35 %	24 %	40 %
ANN	0.45	65 %	43 %	76 %	**ANN**	**0.29**	**48 %**	**36 %**	**53 %**

Level of detail 3					Level of detail 4				
Model	MCC	F-Score	Precision	Recall	Model	MCC	F-Score	Precision	Recall
LR	0.27	41 %	38 %	42 %	LR	0.35	53 %	39 %	58 %
DT	0.32	42 %	43 %	42 %	**DT**	**0.41**	**58 %**	**42 %**	**65 %**
RF	0.22	35 %	34 %	35 %	RF	0.38	54 %	42 %	58 %
GB	0.21	33 %	34 %	33 %	GB	0.34	49 %	40 %	52 %
AdaBoost	0.13	44 %	26 %	44 %	AdaBoost	0.36	57 %	38 %	65 %
ANN	**0.32**	**51 %**	**40 %**	**51 %**	ANN	0.34	50 %	41 %	53 %

models on the level of detail (2) differ only slightly from the models on the level on detail (3), whereas the models on the level of detail (4) differ more strongly from the levels of detail (2) and (3). That emphasizes again an increasing model quality for the regression model with a finer level of detail. Furthermore, there are no outliers in any of the boxplots. Consequently, none of the prediction models within the four levels of detail differs significantly from the other models on the respective level of detail. Nevertheless, the decreasing spread from the level of details (2) to level of detail (4) indicates that with a finer level of detail the model quality of the regression models converges. One possible explanation for the decreasing spread in the regression models is the increasing amount of training data with a finer level of detail—from components to orders to operations—leading to a more solid data base for training the models.

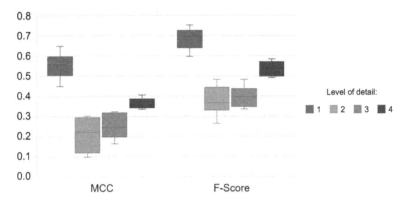

Fig. 5.4 Boxplot of MCC and F-Score for all prediction models on each of the four levels of details

Considering all achieved model qualities, the working hypothesis cannot be confirmed, since—contrary to the working hypothesis—the best result is achieved at the coarsest level of detail (1). Considering the three finer levels of detail only, the model quality is increasing with a finer level of detail, but still, the model quality is below the results on the coarsest level of detail (1).

Consequently, we analyzed possible explanations for the different behavior of the models in our approach regarding their achieved model quality. One possible explanation for the lower model quality on the finer levels of detail (2)-(4) could be, that errors occurring during the prediction of the lead times are cumulated in the postprocessing operations: For each assembly order, n completion dates are predicted according to the number of components. Subsequently, the final output "assembly start delayer" or "no assembly start delayer" is calculated for each component. This calculation of the assembly start delayers is based on the calculated completion dates of all n components and thus includes the errors of all individual calculations of the completion dates. Furthermore, each calculated completion date is composed of a fictitious start date and a predicted lead time.

Both the fictitious start date and the predicted lead time can be subject to errors. In reality, an order can also start on a start date other than the fictitious start date, which can result in a deviation between the predicted completion date and the actual completion date. In summary, the cumulation of errors is one potential explanation for the lower model quality at the finer levels of detail (2)-(4). This explanation is supported, for example, by the authors of [175], who compare simpler with more complex prediction models in their study. Contrary to their initial assumption that under certain conditions more complex prediction models are more accurate, they conclude that simpler models achieve better results. Thus, we recommend for future models for the prediction of assembly start delayers to follow the structure at the coarsest level of detail.

5.6   Section V: Critical reflection of results, limitations, and implications for further research

Missing individual components leading to a delayed assembly start is often an issue for engineer-to-order manufacturers. Thus, the object of consideration in our study tends to be of general nature. Consequently, the case-based research approach as applied research methodology induces legitimate questionability of its comprehensiveness and representativeness for engineer-to-order manufacturers in general. Our results obtained rely on one single exemplary case which might not be representative for all engineer-to-order manufactures. Thus, it might not be generalizable to all cases. Furthermore, due to the defined scope of this study, the considered input features of the prediction models rely on the interviewed experts of the observed company. When transferring the approach to other cases, a new identification of the considered input features might be necessary. Although we tried to overcome these limitations by considering established input features for prediction of lead times in previous studies, further relevant input features might have remained undetected or the considered input features might not be relevant in other cases. Consequently, in future works, the results should be verified with other cases or from a generic point of view. Nevertheless, in the research areas of machine learning and lead time prediction case-based-research is an established research method as it provides necessary training data (see, for example, [70, 71, 126, 128]).

In our case study, the F-scores of all models with a maximum of 75 % and MCC with a maximum of 0.65 were still low and not fully reliable for a practical application. Reasons for not reaching fully reliable ML-models with F-scores close to 100 % and MCCs close to 1.00 are typically a combination of the considered modeling approach, the ML-algorithm, and the data base [137, 163–165]. To reach the current values of our best F-score and MCC we applied four different modeling approaches and several different ML-algorithms including different structures of the ANN and tuning of the hyperparameters. Thus, we infer, that neither a further optimization of the modeling approach nor the ML-algorithm used leads to a significant improvement of the model quality. One possibility to further improve the model quality could be to enhance the data basis used for training the models, as the data base also has an essential influence on the model quality [137, 138]. In manufacturing processes, especially at machine and plant manufacturers, there are typically many reasons for a delay such as missing raw material, problems when setting up the machine, machine downtimes, issues during the execution of an operation, rework, quality problems with a certain material, or also nonproduction related reasons such as issues in a global supply chain or even the weather (see, for example, [64, 176, 177]). Thus, to ensure a fully reliable model, all the potential disruptions would need to be considered in the machine learning model, and consequently, the data base needs to cover all that information about the respective disruptions as well. In our case study, with a selection of all available order data, machine data, material data and system information, we cover a subset of all information about potential disruptions only. Accordingly, we assume that considering further information about typical disruptions occurring at the exemplary chosen machine and plant manufacturer such as detailed information about the production process at the raw material supplier or maintenance data, could significantly improve the model quality. Consequently, we encourage further studies to consider additional data fields about potential disruptions when setting up a model predicting assembly start delayers to further optimize the model. Without an improvement of further approx. 15–20 % in F-score, the model will not be suitably usable for manufacturing companies. Nevertheless, our study is a good starting point in the research area of predicting assembly start delayers analyzing essential basics regarding the modeling approach for future studies.

A further area for future work could be the provision of background information on the identified assembly start delayers. The current models are only able to identify the assembly start delayers. However, there is no information on the reasons for the occurrence of an identified assembly start delayer given that would explain why the component was supplied late. In order to be able to prevent a potential assembly start delayer by defining suitable counter measures, information about the causes of the delay is of immense importance. Thus, the investigation of how methods from the area of explainable AI can support the provision of background information in the prediction of assembly start delayers could be a potential further research field.

Besides the considered limitations and implications for further research, we could satisfactorily answer our initially stated research question "how does the level of detail of the modelling affect the model quality to predict assembly start delayers". In our case study, we could show, that the level of detail of the modelling significantly affects the model quality. The best modelling approach in our case study was to apply a classification model to predict assembly start delayers. Thus, the target of our study was achieved.

5.7 Section VI: Conclusion

Adherence to delivery dates is a decisive factor for manufacturing companies to assert themselves in globalized markets. A central aspect to meet delivery dates is an assembly of a product on time. Delays in the processes upstream of the assembly such as the in-house production of individual components can have a negative effect on the adherence to delivery dates. In order to prevent delays in the processes upstream of the assembly, in this work a supervised learning model to predict missing components for the assembly start, so-called assembly start delayers, in early phases of the production process was developed. Here we analyzed the level of detail of the prediction model since it can have a significant impact on the model quality. An increase in the level of detail usually leads to a higher model accuracy, but with a degressive characteristic [165]. Thus, we formulated the following working hypothesis: "The model quality for the prediction of assembly start delayers increases with a finer level of detail." In order to verify the working hypothesis, in total 24 ML-models were created, which differ in their level of detail and the utilized ML-algorithm, but with the prediction of assembly start delayers as their common target. Here a case-based research approach was applied. As an exemplary case for this research approach, a machine and plant manufacturer was chosen and real-world data was applied.

The model architectures of the models on the four levels of detail are different. The models on the coarsest level of detail predict assembly start delayers utilizing a classification approach. The models on the three finer levels of detail predict assembly start delayers based on a prior lead time prediction via a regression and subsequent postprocessing operations. The regression models differ in the lead times considered. A finer level of detail corresponds to a finer consideration of the lead time. Specifically, the component, order and operation lead times were considered. In the subsequent postprocessing operations, the assembly start delayers were identified based on the predicted lead times. Finally, the output of all 24 prediction models on the four levels of detail was the binary classification "assembly start delayer" or "no assembly start delayer" for every component. To evaluate the model quality of all 24 models a confusion matrix was created and the metrics MCC, F-score, precision and recall were calculated.

The comparison of the model qualities at the four levels of detail showed that, contrary to the working hypothesis, the model on coarsest level of detail— the classification approach—had the best model quality. In contrast, an increase in model quality with a finer level of detail was evident within the regression models. In our study, in total, a finer level of detail did not lead to the best result obtained. Consequently, the working hypothesis could not be confirmed. As a possible explanation for the lower model quality on the three finer levels of detail we identified a cumulation of errors occurring during the prediction of the lead times in the postprocessing operations.

In total, we successfully implemented 24 ML-models to predict assembly start delayers and gave insights in the performance of different modeling approaches. Such prediction models can be useful to identify assembly start delayers in early phases of the manufacturing process and to enhance the delivery performance machine and plant manufactures if a sufficiently high model quality is achieved.

5.8   Appendix

The hyperparameter used in the prediction models were optimized utilizing a grid search and cross validation algorithms (GridSearchCV) from scikit learn. Table 5.5 to Table 5.8 summarize the utilized hyperparameters in the different models on the four levels of detail.

Table 5.5 Hyperparameters of the prediction models on level of detail 1

Model	Hyperparameters					
	c	penalty	loss	max iter	dual	n estimators
SVM	1	l2	hinge	1000	True	-
	min samples split	min samples leaf	max features	max depth	learning rate	n estimators
DT	2	1	None	None	-	-
RF	2	2	70	None	-	500
GB	15	6	80	7	0.2	2000
AdaBoost	30	15	50	3	1	800
	momentum	weight decay	dampening	epochs	learning rate	
ANN	0.9	0	0	450	0.01	

Table 5.6 Hyperparameters of the prediction models on level of detail 2

Model	Hyperparameters			
	normalize	fit intercept	copy x	n_jobs
LR	False	True	True	None

	min sam- ples split	min sam- ples leaf	max fea- tures	max depth	learning rate	n estima- tors
DT	12	9	10	10	-	-
RF	12	8	auto	10	-	60
GB	2	3	10	7	0.8	10
AdaBoost	25	12	22	5	0.7	30

	verbose	batch size	epochs
ANN	1	64	450

Table 5.7 Hyperparameters of the prediction models on level of detail 3

Model	Hyperparameters			
	normalize	fit intercept	copy x	n_jobs
LR	False	True	True	None

	min samples split	min samples leaf	max features	max depth	learning rate	n estimators
DT	5	1	40	5	-	-
RF	10	9	auto	None	-	140
GB	2	3	None	2	0.5	10
AdaBoost	5	2	30	7	10	90

	verbose	batch size	epochs
ANN	1	64	500

Table 5.8 Hyperparameters of the prediction models on level of detail 4

Model	Hyperparameters					
	normalize	fit intercept	copy x	n_jobs		
LR	False	True	True	None		
	min samples split	min samples leaf	max features	max depth	learning rate	n estimators
DT	2	7	30	None	-	-
RF	10	3	auto	None	-	100
GB	8	7	None	7	2	10
AdaBoost	2	8	33	None	2	40
	verbose	batch size	epochs			
ANN	0	32	600			

5.9 Presentation of own contribution

All the work presented within this publication was carried out by me. This includes the definition of research questions and hypotheses, the definition of the research approach including the set up to compare the performance of ML-models on different levels of details and different ML-algorithms on each level, the collection, cleaning and aggregation of data from the manufacturing company under consideration, the feature engineering and setting up of the data model, the training and tuning of all ML-models including writing of the respective code in Python as well as the evaluation and critical reflection of the results. Further, transforming the work into text and writing of the publication were entirely in my hands. All three co-authors Prof. Dr. Burggräf, Dr. Wagner and Mr. Heinbach contributed with ideas to the research concept and in the internal review process.

Publication IV: Predicting Supplier Delays Utilizing Machine Learning—a Case Study in German Manufacturing Industry

6

F. Steinberg, P. Burggräf, J. Wagner, B. Heinbach, T. Saßmannshausen, and A. Brintrup, "A Novel Machine Learning Model for Predicting Late Supplier Deliveries of Low-Volume-High-Variety Products with Application in a German Machinery Industry," *Supply Chain Analytics*, p. 100003, 2023, https://doi.org/10.1016/j.sca.2023.100003.

6.1   Abstract

Although Machine Learning (ML) in supply chain management (SCM) has become a popular topic, predictive uses of ML in SCM remain an understudied area. A specific area that needs further attention is the prediction of late deliveries by suppliers. Recent approaches showed promising results but remained limited in their use of classification algorithms and struggled with the curse of dimensionality, making them less applicable to low-volume-high-variety production settings. In this paper, we show that a prediction model using a regression algorithm is capable to predict the severity of late deliveries of suppliers in a representative case study of a low-volume-high-variety machinery manufacturer. Here, a detailed understanding of the manufacturer's procurement process is built, relevant features are identified, and different ML algorithms are compared. In detail, our approach provides three key contributions: First, we develop an ML-based regression model predicting the severity of late deliveries by suppliers. Second, we demonstrate that prediction within the earlier phases of the purchasing process

F. Steinberg, *Machine Learning-based Prediction of Missing Parts for Assembly*, Findings from Production Management Research ,
https://doi.org/10.1007/978-3-658-45033-5_6

is possible. Third, we show that there is no need to reduce the dimensionality of high-dimensional input features. Nevertheless, our approach has scope for improvement. The inclusion of information such as component identifiers may improve the prediction quality.

6.2   Section I: Introduction

Today, companies source their goods from all over the world, with multi-modal transport chains delivering everything from simple components to highly complex products. Efficient procurement of goods is largely dependent on supplier performance. Goods that are delivered with insufficient quality, quantity, or with a time delay, lead to disruptions in manufacturing companies that need these goods in their production. Those disruptions are especially serious for manufacturers whose value stream is critically dependent on the assembly process, where several material flows from different suppliers and the in-house manufacturing converge [46]. Here, only a single missing component can impede the timely start of the entire assembly process involving up to several thousand components [52]. Thus, to ensure a timely start of the assembly process, and consequently, to meet their delivery dates, for manufacturing companies it would be helpful to predict potential delays in their upstream supply chain.

With the advances in data analytics utilizing machine learning (ML) and the availability of large-scale, unstructured data sets, ML-based prediction models are becoming more and more established. Despite these advances, in supply chain management (SCM) recent review articles have identified a predominance of descriptive analytics rather than predictive analytics, except for demand forecasting [179–181]. As the complexity of supply chains is continuously increasing, predicting supply chain disruptions before they occur is becoming increasingly important as well [182]. Recent developments in digitalization technologies, such as the internet of things or artificial intelligence, present novel opportunities for predicting disruptions in SCM [183, 184]. However, contributions presenting such prediction models are limited in SCM literature [179–181].

In the body of literature of the specific area of identifying and quantifying late deliveries of suppliers, currently, there are only two approaches available: First, the research of Brintrup et al. [20], and second, the research of Baryannis et al. [21]. Both approaches are limited in their use of classification algorithms and struggle with the curse of dimensionality, making them less applicable to low volume high variety manufacturing settings. In addition, the prediction models are based on information that is available after the order has been placed. In both approaches, we see the following three shortcomings, which we want to address in our manuscript: First, unlike a classification model, a regression model can make predictions of delivery delays in calendar days, thus assessing the severity of a delay and providing a more valuable prediction. In real-world manufacturing applications, the evaluation of the severity of a delay is essential to select and prioritize appropriate countermeasures. Nevertheless, a regression model predicting delivery delays of orders placed at suppliers of low volume high variety manufacturers is currently not available in the body of literature. Second, the time of a prediction is essential for the implementation of potential countermeasures—in our work, we define the time of prediction as the point of time in the purchasing process when the prediction model is applied, and the prediction is made; this could be, for example, the time of creating an order request, the time of placing an order, or time of receiving a delivery confirmation. An earlier prediction than at the time of placing an order as used by Brintrup et al. [20] and Baryannis et al. [21], such as after the creation of an internal purchasing request, would provide more opportunities for countermeasures in case of a predicted delay. However, typically, less information is available at earlier time points, which can cause prediction models to have lower quality. But, in the specific area of predicting late deliveries of suppliers, there has been no study of the influence of the time of the prediction on the model quality. Third, both available approaches limit the scope—e.g., they exclude components that are ordered less than five times—because they struggle with the curse of dimensionality. Especially in low volume high variety environments, however, it is typical that a large proportion of the components are designed individually for each customer need and are therefore procured only once. Such a restriction in the scope would then lead to a large proportion of the required components no longer being considered in the prediction model, which in turn limits the practical applicability of the model.

Reflecting on the shortcomings of the above-mentioned approaches, in this work we focus on the following three research questions (RQ):

- RQ1: Are regression algorithms capable to predict delivery delays of orders placed at suppliers of a low volume high variety manufacturer?
- RQ2: What is the impact of the time of the prediction on the model quality?
- RQ3: Is the curse of dimensionality an issue when setting up a regression model predicting delivery delays of orders placed at suppliers of a low volume high variety manufacturer?

To answer these research questions, we conduct a case study at a machinery manufacturer focusing on the prediction of potential delays of components delivery dates ordered at suppliers. A case-based research approach is an objective, detailed investigation of a current phenomenon where the researcher has little control over real events [170]. One motivation for the case-based research approach is to gain insights for real needs of manufacturing companies, rather than to develop theories without practical relevance [145]. Furthermore, a case-based research approach has already been successfully applied in the area of predicting delivery delays [20, 21, 181]. Accordingly, a case-based research approach is an appropriate method to answer the research question and to investigate the working hypothesis.

In detail, we set up a two-stage experimental plan for this purpose. In the first stage, the quality of ML-based regression models predicting delivery delays is compared at different times of the prediction and thus the influence of the time of the prediction on the regression model quality is quantified. At each time of the prediction, a range of standard ML algorithms is applied to an identical data set to allow comparability between prediction time points and algorithms. With this comparison, we can also evaluate if regressions algorithms are capable to predict delivery delays. Thus, in the first stage of the experimental plan, we will answer RQ1 and RQ2. In the second stage, we then investigate the impact of reducing the dimensionality of high dimensional input features on the model quality by comparing different exclusion criteria and thus answering RQ3. The setup of all ML models within the experimental plan follows the established procedure model Cross Industry Standard Process for Data Mining (CRISP-DM) [146, 147] consisting of the six phases business understanding, data understanding, data preparation, modelling, evaluation, and deployment. To ensure comparability across our experimental plan, the datasets, features, and ML algorithms utilized remain equal in both stages of the experimental plan.

The evaluation of the case study provides the following three main contributions:

- We show that an ML-based regression model can predict delivery delays of orders placed at suppliers of machinery manufacturing in calendar days.
- We demonstrate that an early prediction within the purchasing process based on information available after creating the internal order request is possible and only slightly worse than a later prediction.
- We show that there is no need to reduce the dimensionality of high dimensional input features.

This paper is structured as follows. First, section 2 introduces the state of the art in predictive data analytics in SCM with a focus on available approaches predicting delays of orders. Section 3 is structured according to the CRISP-DM framework giving a description of the case study dataset and details about the feature selection and engineering as well as the setup of the experimental plan, the ML models and the results. Further, a comparison of our model performance with recent approaches is conducted. Subsequently, Section 4 critically reviews the limitations of our approach and the results obtained. Furthermore, the implications for further research are derived. Finally, a summary is given in the last section.

6.3   Section II: State of the art: Predictive Analytics in Supply Chain Management

To reduce the impact of a disruption, there are typically two options. First, to minimize its risk of occurrence, and second to strive for a resilient supply chain that quickly returns to its original state after a disruption [185–187]. These two options are covered by two individual domains in SCM, namely, supply chain risk management (SCRM) and supply chain resilience. In both and the superordinate field SCM as well, data analytics is one of the core tools used. Waller and Fawcett [188] define the term data analytics in SCM as '*the application of quantitative and qualitative methods from a variety of disciplines in combination with SCM theory to solve relevant SCM problems and predict outcomes, taking into account data quality and availability issues*'. They further classify predictive analytics as a subset of data analytics to improve supply chains and mitigate risks by forecasting what could probably happen in the future. In contrast, Wang et al. [179] and Nguyen et al. [180] give a wider differentiation of data analytics in SCM. They classify the available approaches in descriptive, predictive, and prescriptive analytics. Descriptive analytics in SCM focuses on what happened in the past (see, for example, [189–191]). Predictive analytics attempts to predict and explain events that will occur in the future (see, for example, [192–194]). Prescriptive analytics is using data and algorithms to find alternative decision options (see, for example, [195–197]). Out of these three categories, current research focuses mainly on prescriptive analytics rather than descriptive and predictive analytics [180]. Nevertheless, as typical for data analytics in general and not just in the area of SCM, the performance of prescriptive models relies on descriptive and predictive models [179, 180]. Thus, the aforementioned review papers call for new research in descriptive and predictive analytics in SCM. Hence, we contribute to the body of literature with a case study focusing on predictive analytics in SCM.

Bienhaus and Haddud [198], Ivanov et al. [199], and Queiroz et al. [200] highlight the premise and use of big data and Artificial Intelligence in the digital transformation of the procurement process as one of the key factors to enhance the competitiveness, efficiency and profitability of companies' supply chains. With the continuously increasing availability of a wider volume, velocity, and variety of data, new opportunities arise to revolutionize the impact of data analytics approaches [188, 201, 202].

In SCM as a whole, ML and other data mining techniques are frequently considered for demand forecasting (see for example, [203–205]), determining retail prices in supply chains including the handling of financial flows (see, for example, [206–208]), or dealing with the effects of low-frequency high-impact disruptions on supply chains such as the COVID-19 pandemic (see, for example, [209–212]). In the specific subdomains of procurement and logistics current research mainly supports the selection of potential suppliers for specific products (see, for example, [213–215]) or deals with problems of vehicle routing [180, 216], but missing material due to late deliveries is a neglected area of research [177]. Models predicting late deliveries of suppliers are still rare. To the best of our knowledge, we identified only two articles focusing on predicting late deliveries of suppliers based on real data sets using machine learning.

Baryannis et al. [21] proposed an ML-based approach predicting late deliveries of suppliers with a focus on their interpretability to be able to support decision-making following the prediction. Given a real data set of a multi-tier aerospace manufacturing supply chain consisting of product data such as the part number and price, order data such as due dates, quantities ordered and original delivery requests, and delivery data such as receipt date and quantity receipt they compare the performance and interpretability of support vector machines (SVM) with decision trees (DT). Prioritizing interpretability over performance they recommended DT as the ML algorithm of choice resulting in slightly worse performance metrics. While we agree with the need for more interpretable ML in SCM, we postulate that other algorithms such as ensemble algorithms or ANN which suffer from interpretability also need to be studied to be able to present the full range of options to the decision-maker.

Brintrup et al. [20] presented a case study at an original equipment manufacturer (OEM) predicting delivery delays of Tier 1 suppliers also based on historical product data such as product description and product type, order data such as order date and supplier ID, and delivery records such as the received date. Comparing five ML algorithms they identified a random forest algorithm (RF) outperforming SVM, logistic regression, linear regression, and k-nearest neighbour algorithm. Similar to Baryannis et al. [21], more complex ML algorithms such as ANN might have performed better but remained uncovered.

Further, one of their biggest challenges was the curse of dimensionality due to high variability in their categorical features leading to a high number of variables in their feature space. To reduce the dimensionality, they excluded data with less than five samples for each categorical attribute. Thus, they excluded for instance suppliers who delivered less than five times or components that were ordered less than five times. This restriction in the variability of the input data might be a limitation when transferring the approach to industries that focus on low volume high variety customized production, where components ordered at suppliers may vary strongly. Thus, we postulate that the impact of such a limitation on models predicting delivery delays in low volume high variety production needs to be investigated.

In summary, there are several approaches available focusing on data analytics in supply chain management. However predictive analytics in SCM remains an understudied topic. A specific area that needs further attention is the identification and quantification of late deliveries. Extant approaches are limited in their use of classification algorithms and struggle with the curse of dimensionality, making them less applicable to low volume high variety settings.

Hence, we contribute to the body of literature with a novel case study in predictive analytics in supply chain management using machine learning—specifically in predicting delivery delays of suppliers with a supervised learning approach using a real data set from a machinery manufacturer. Here, we compare simple ML algorithms such as DT, RF, or SVM with more sophisticated approaches such as ANN. Furthermore, we analyse the impact of reducing the dimensionality of high dimensional input features on the model quality and the practical usability of the model by comparing different exclusion criteria.

6.4   Section III: Case Study

For our case study, we choose an OEM in the German machinery manufacturing industry that builds complex products made up of several thousand individual components and several hundred sub-systems. The products are typically individually designed for customers' needs. The upstream supply chain consists of Tier 1 suppliers for finished components that are used directly in assembly, as well as Tier 1 suppliers for raw materials that are then mechanically processed in the company's production facilities. Approximately 80 % of the total number

Fig. 6.1 Procedure of the CRISP-DM framework applied in the case study [146, 147, 217]

of components are purchased as finished components and 20 % are processed within the company. Most of the raw material supply for in-house production is decoupled via storage with suitable safety stocks. Further, in-house production is set-up to be flexible so that potential delays in raw material supply can be partially compensated. In contrast, most of the finished components are ordered individually for each customer project, so delays directly influence the timely start of the assembly. Thus, in our case study, we focus on the late deliveries of finished components.

To answer the research questions, we set up a two-stage experimental plan. In stage one we quantify the impact of the time of the prediction on the quality of ML-based regression models predicting late deliveries and simultaneously analyse if an ML-based regression model is capable to predict delivery delays in calendar days. In stage two we evaluate if there is a need to limit the scope to overcome the curse of dimensionality when predicting late deliveries with ML-based regression models within our exemplary case. To develop the different ML models we applied the established CRISP-DM procedure model (see Fig. 6.1) [146, 147]. Further, to ensure comparability across our experimental plan, the datasets, features, and ML algorithms utilized remain equal in both stages of the experimental plan. Thus, the general process to develop the different ML models remain equal as well. Consequently, the next subsections are following the phases of CRISP-DM business understanding, data understanding, data preparation, modelling, evaluation, and deployment. As we mainly focused on the development of the model, we excluded the last phase Deployment.

6.4.1   Business Understanding

The Business understanding phase typically includes a description of the business problem and a transfer of the business problem into concrete requirements and objectives for further data analysis. Thus, the first phase of the CRISP-DM provides a central basis for all the following steps and decisions in the data mining process.

As the business problem is delays in the assembly process due to delayed deliveries of finished components, the objective from a business perspective was to prevent these delays. Predicting potentially delayed deliveries of finished components ordered at the suppliers can support the OEM's purchasing department to take countermeasures such as speeding up the supplier's manufacturing process, choosing a different means of transport for the delivery, or even choosing a different supplier. The OEM's business process itself is typical for a machinery manufacturer. The relevant components for assembly are first defined in a design and material planning process. Then, after a make-or-buy decision, either a production order for in-house production or a purchasing request is created in the company's Enterprise Resource Planning (ERP) system for each required component. The purchasing request then initiates the purchasing process following the established standards. Here, we wanted to support the purchasing process with a prediction of potentially delayed deliveries as early as possible. Together with the domain experts at the OEM, we identified three potential times of the prediction, which is defined as the point in time within a procurement process a prediction model is applied:

1. *Purchasing request:* A prediction based on all information available after creating the purchasing request in the ERP System.
2. *Order placement*: A prediction immediately after placing an order at a selected supplier.
3. *Delivery confirmation*: A prediction immediately after receiving a delivery confirmation of the supplier including the confirmed delivery date.

Although it may seem trivial that a prediction at a later point in time might be of higher quality due to more available information about the purchasing process, an early prediction would be more helpful to take effective countermeasures. Here, one objective of our case study is to identify a good trade-off between the time of the prediction and the prediction quality.

Further, the type of prediction is interesting as well. Binary classification of the deliveries in *late* and *in time* as applied by Brintrup et al. [20] and Baryannis et al. [21] is less valuable than a regression model predicting potential delays in working days. A prediction in working days would give additional information about the severity of a delay. Thus, a regression was our modelling approach of choice. Here, we applied several ML algorithms such as tree-based algorithms, support vector machines, or neural networks utilizing the Scikit-learn library in Python.

Hence, we transformed the business problem—missing components at the start of assembly—into a machine learning problem—predicting delivery delays in the supply of externally purchased finished components utilizing a supervised learning approach. To compare the prediction quality of the different ML models to be set up in the modelling phase first defined metrics for each prediction model. More specifically, we selected the following established metrics for regression models as our metrics of choice: mean absolute error (MAE), root mean squared error (RMSE), and coefficient of determination (R^2). Here MAE and RMSE are defined as

$$MAE = \frac{1}{n} \sum_{i=1}^{n} |y_{i,true} - y_{i,predicted}| \qquad (6.1)$$

$$RSME = \sqrt{\frac{1}{n} \sum_{i=1}^{n} (y_{i,true} - y_{i,predicted})^2} \qquad (6.2)$$

where n is the number of samples considered, $y_{i,true}$ is the actual value, and $y_{i,predicted}$ is the predicted value. R^2 is defined as

$$R^2 = 1 - \frac{\sum_{i=1}^{n} (y_{i,true} - y_{i,predicted})^2}{\sum_{i=1}^{n} (y_{i,true} - \bar{y}_{true})^2} \qquad (6.3)$$

where $\bar{y}_{true}$ is the empirical mean of all y_{true}. Further details about the equations can be looked up in [218, 219].

6.4.2   Data Understanding

In the second phase, data understanding, following Wirth and Hipp [147], we collected and analysed the data to develop a solid understanding of the dataset. For our prediction model, we included data from two business processes of the company under consideration. First, the material planning process was collected, which included information about the bill of material (BOM) and demand dates of components for processing in assembly based on backward scheduling. Second, the purchasing process containing information about the orders including the respective deliveries was collected. As both processes are executed and documented in the company's ERP system, this was also our data source of choice. The data export with a period under review of three years consisted of two separate CSV-files containing purchasing orders and BOM items. With one ID field, we were able to merge the two separate files. When merging the data, since predicting delivery delays of the purchasing orders was our target, we kept the orders as our object under consideration and expanded them with information from the BOM. The total data set comprised 119,610 purchasing orders with information from 17 different data fields (see Table 6.1). As the dataset is confidential, we are not allowed to make it available to third parties and cannot publish it.

As target variable for our prediction model and thus, to predict potential delivery delays we calculated a delivery date lateness (DDL) considering the actual delivery date and the demand date. In detail, we utilized the formula

$$DDL = Delivery\ date - Demand\ date \tag{6.4}$$

to calculate the DDL. Here, a negative DDL indicates a delivery before the demand date and a positive DDL indicates a delayed delivery. Thus, on the one hand, companies can predict the severity of delivery delays in calendar days for late deliveries. On the other hand, with the prediction of the duration components are delivered before the demand date, companies can allocate space for inventory and calculate days of inventory turnover predictably.

Next, we performed an exploratory data analysis to understand the main characteristics of our dataset using statistical graphics such as boxplots, scatter plots, and histograms. Here, we first analysed the distribution of the DDL as our target variable for our prediction model (see Fig. 6.2). It is noticeable that with a portion of approx. 86 % of most of the orders were delivered before or in time to the demand date, and with a portion of approx. 14 % only a few orders were delivered delayed. Thus, only a slight portion of all orders is the main reason for missing material in the assembly. Further, it is noticeable that few orders are delivered several months or even up approx. 0.75 years before the demand date. Potential reasons for these high DDLs are shifts of the assembly after the actual delivery date due to shifts in the customer's order or an assignment of material from a previous purchasing order, that has been placed in stock, to a different assembly order. Nevertheless, we included these purchasing orders in our prediction model, as these were real and not data errors.

Table 6.1 Overview of data fields

Data field	Data format	Description
Product description	Text	Short description of the component
Drawing number	Alphanumeric	Unique drawing identification
Order quantity	Integer	Number of components ordered
BOM item created	Date	When the material planning process is completed
Demand date	Date	When the component is required in the assembly
Order created	Date	When purchasing process was initiated
Order date	Date	When the order was placed
Requested delivery date	Date	Requested delivery date of the purchasing department
Confirmed date	Date	Confirmed delivery date of the supplier
Confirmation received	Date	When the supplier confirmed the delivery
Delivery date	Date	When did the supplier deliver
Order-ID	Integer	Purchase order number
Order method	Alphanumeric	Category indicating how the order was placed
Supplier-ID	Integer	Unique supplier identification
Supplier	Text	Legal name of the supplier
Material	Text	Short description of the material (e.g., S235)
Gross weight	Integer	Gross weight of the component

Moreover, we analysed the product portfolio based on CAD drawing numbers (see Fig. 6.3) and supplier structure based on supplier-IDs (see Fig. 6.4) to get a better understanding of the variety of categorical data fields in our data set. This analysis was also an indicator of the curse of dimensionality [220] that might affect our prediction model. Looking at the product structure, a portion of 61 %

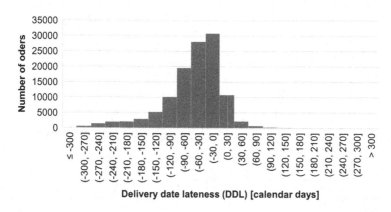

Fig. 6.2 Distribution of the delivery date lateness (DDL)

of all components were ordered twice or less, whereas with a portion of 4 % some components were ordered 20 or more times. Additionally, the components ordered twice or less comprised 17 % of all orders placed and the components ordered 20 or more times comprised 30 % of all orders placed. Therefore, limiting the scope of consideration to overcome the curse of dimensionality to, for example, components that have been ordered at least twice would mean that 61 % of all components and thus 17 % of all orders would not be considered in the prediction model. Other authors such as the authors in [20] limit their scope even stricter to components that have been ordered at least five times. Nevertheless, excluding 61 % of all components, when limiting the scope to components that have been ordered at least twice was a non-negligible limitation in our case study. Consequently, within our modelling phase, we additionally analysed the trade-off between the scope limitation in the variety of components and the achievable model quality. In detail, we quantified the impact of such a limitation on the model quality by comparing prediction models with different exclusion criteria.

Looking at the supplier structure, in addition to the product portfolio, it was noticeable that 22 % of all suppliers received a maximum of three orders, which, however, only accounted for 0.2 % of all orders. On the other hand, 0.3 % of all suppliers received 10,000 each or more orders and accounted for 22 % of all orders. Here, limiting the scope to suppliers who received more than three orders seemed reasonable, since only 0.2 % of all orders would have been excluded. Thus, a limitation within the supplier structure could help to overcome the curse

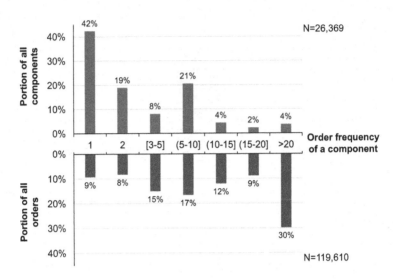

Fig. 6.3 Distribution of the product portfolio

of dimensionality excluding only a tiny portion of all orders. Nevertheless, further analysis showed, that 60 % of the respective orders placed at suppliers that received three or fewer orders were delivered too late. Comparing this portion of late deliveries with the overall portion of too late deliveries, which was 14 %, revealed an outstanding potential of late deliveries of suppliers who received three or fewer orders. Thus, even this seemingly negligible limitation in the supplier structure would have meant that suppliers with outstanding potential for disruption would have been excluded from the scope, which was another non-negligible limitation in our case study. Consequently, also in the supplier structure, we analysed the trade-off between the scope limitation in the variety of suppliers and the achievable model quality by comparing prediction models with different exclusion criteria.

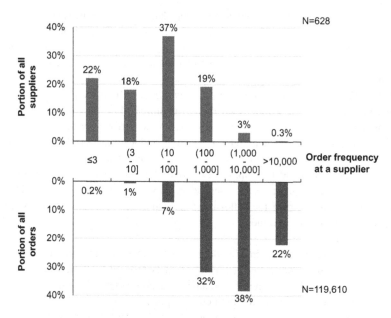

Fig. 6.4 Distribution of the supplier structure

6.4.3 Data Preparation

With the understanding gained about the dataset, we then started to prepare the final dataset for our prediction models. First, we looked at the features in date-format. Using a date as an input feature for a supervised learning approach means the prediction model is trained on those historical dates as well. In our case, whenever the model will be used in the future, the input dates would be in a different year and not comparable to the dates the model was trained on. Thus, we transformed the features into a date-format. Instead of the actual date as one feature, we used four features: The first three were the year, month, and day of the date in integer-format, and the fourth was the deviation of the respective date to the demand date also in integer-format.

Next, following Kuhn and Johnson [151], we performed a correlation analysis to identify the relevant features for our prediction model. For correlation between categorical features, we calculated Cramér's V, and for correlation between continuous variables, we calculated the Pearson correlation coefficient. The interpretation of the correlation coefficient followed the definitions given by Akoglu [221]. First, we analysed the correlation between the target variable DDL and our input features revealing only weak correlations of the date fields in the first three transformations year, month, and day. Instead, the fourth transformation of all date fields, which was the deviation of the respective date to the demand date, showed strong correlations. Thus, we excluded the transformations year, month, and day and only kept the transformation into the deviation to the demand date. Further, we analysed the correlation within our input features to identify feature dependencies. This analysis showed a strong correlation between the drawing number and the product description, and a strong correlation between the supplier-ID and the supplier in text format. Consequently, we excluded the product description and the supplier in text format, and only kept the drawing number and supplier-ID to reduce dependencies within our input features. Thus, in summary, we selected the following twelve independent variables to be the input features for our ML models:

1. *Drawing Number:* The drawing number is a unique identifier of a component allowing one to determine if a component is ordered multiple times. Thus, based on the drawing number, delays of a component ordered in the past can be used as a base for predicting a potential delivery delay in future. Consequently, this variable is important for predicting delivery delays.
2. *Order quantity:* The lead time of a production lot typically scales with the number of components to be manufactured. Consequently, the number of ordered components influences the delivery times of suppliers and is thus important for the prediction of delivery date delays.
3. *Material:* The idea behind using the material as an independent variable is that certain materials from which a component is made could be delivered more quickly than rare materials. Therefore, the use of the material could reveal patterns that have an influence on the prediction of delivery date delays.
4. *Gross weight:* The effort required to handle and transport components depends on the weight of a component. Further, the weight of a component could be an indicator of the effort required to manufacture it. Thus, the usage of the gross weight of a component is a relevant variable for predicting delivery delays.
5. *BOM-item created:* The creation of the BOM item is the first event at which the need for a component is systemically recorded and thus known. At the company under consideration, the creation of demand is the responsibility of the design department. Without that demand, the purchasing process is not

initiated. Thus, the duration between the date of creating the BOM item and the demand date is relevant for predicting delivery delays.

6. *Order created:* After the BOM item is created and after the decision has been made to procure the component, a purchasing order is created and thus, the purchasing process is initiated. Consequently, delivery delays depend on the available duration between order creation and the demand date.

7. *Order date:* The next step in the purchasing process is ordering the component from a supplier. The duration between the order date and the demand date represents the maximum delivery time for a supplier to deliver on time. Thus, the order date has an influence on potential delivery delays.

8. *Request delivery date:* With the order of a component, a requested delivery date is submitted to the supplier. Here, the respective purchaser can make manual adjustments to the demand date—the purchaser can request a delivery date before, contemporary or after the demand date. Consequently, the requested delivery date is essential when predicting delivery delays.

9. *Order method:* When placing orders, the company has different levels of atomization and different communication protocols to their supplier—e.g., Electronic Data Interchange (EDI), E-Mail, or phone calls. This might have an influence on the delivery times and thus on predicting delivery delays.

10. *Supplier-ID:* The supplier ID is a unique identifier of the supplier allowing us to determine if several components are ordered from the same supplier. Thus, based on the supplier ID, delays of deliveries of the respective supplier in the past can be used as a base for predicting a potential delivery delay in future. Consequently, this variable is important for predicting delivery delays.

11. *Confirmed date:* With the confirmation of the order, the supplier provides an estimated delivery date. This information is important for predicting delivery delays.

12. *Confirmation received:* The duration between ordering and receiving a confirmation from a supplier might be an indicator of suppliers' available capacities and delivery performance. Thus, this information might also indicate delivery delays.

Next, since tree-based classifiers and neural networks from the Scikit-learn library can only be trained on numerical variables in Python [154], we needed to convert our categorical variables into numerical variables. One common and established method for this conversion is One-Hot-Encoding. But, without any limitation in the components and suppliers considered One-Hot-Encoding of only these categorical features would have increased the number of dimensions by 26,997 resulting in a high dimensional sparse matrix. To avoid potential memory and

Table 6.2 Overview of the transformed data set for the prediction models

Data field	Time of the prediction		
	1. Purchasing Request	2. Order Placement	3. Delivery Confirmation
Drawing number (Binary-Encoded to 15 features)	X	X	X
Order quantity	X	X	X
Material (Binary Encoded to 9 features)	X	X	X
Gross weight	X	X	X
BOM-item created (Delta to demand date)	X	X	X
Order created (Delta to demand date)	X	X	X
Order date (Delta to demand date)		X	X
Requested delivery date (Delta to demand date)		X	X
Order method (Binary Encoded to 3 features)		X	X
Supplier-ID (Binary Encoded to 10 features)		X	X
Confirmed date (Delta to demand date)			X
Confirmation received (Delta to demand date)			X

computability concerns for our prediction models, and as we wanted to avoid a limitation of our scope, we followed the recommendation of Seger [222] and performed Binary-Encoding instead of One-Hot-Encoding resulting in an increase in the number of dimensions by only 25 without limiting the scope. Nevertheless, we still analysed the impact of a limitation in the scope on the model quality within our modelling phase (see section III.4).

Finally, after further data pre-processing operations such as discretization and normalization (see, for details, [152, 153]), we finalized the data set for our prediction models. In total, we transformed the initial 17 data fields into 45 features through data pre-processing (see Table 6.2). As one of our data mining targets was to analyse the trade-off between the prediction quality and the time of the prediction in terms of the three times 'purchasing request', 'order placement' and 'delivery confirmation' we assigned the 45 features to the different times of the prediction. After defining the input features for the different prediction models the Data Preparation was completed.

6.4.4   Modelling

After understanding the available data and defining the features of our ML models, we set up an experimental plan (see Table 6.3) to give a quantified response to the objectives of our study—answering the research questions. In this phase of the CRISP-DM procedure, we only define the experimental plan to answer the research question. The execution of the experimental plan—the training and evaluation of the ML models—is considered in the results phase.

The experimental plan consisted of two stages. The first stage focused on the trade-off between the time of the prediction and the model quality utilizing different ML algorithms. Simultaneously, as we set up regression models in this first stage, we can also evaluate if regressions algorithms are capable to predict delivery delays. Thus, in the first stage of the experimental plan, we can answer RQ1 and RQ2. The second stage is about analysing the impact of a limitation in the scope in terms of a limitation in the product portfolio and supplier structure on the model quality and thus, is designed to answer RQ3.

In detail, in the first stage, we plan to compare different regression models for each of the three times of the prediction 'purchasing request', 'order placement', and 'delivery confirmation' in the experimental plan. The models in each of the different times of the prediction differentiate in the ML algorithm used. In detail, we compared the performance of a Linear regressor (LR), a Support Vector regressor (SVR), a Decision Tree (DT) regressor, a Random Forest (RF)

Table 6.3 Overview of the experimental plan

ML algorithm	Stage 1—Time of the prediction			Stage 2—Limitation in the scope		
	Purchasing Request	Order Placement	Delivery Confirmation	Five or more orders	Two or more order	No limitation
LR	X	X	X	Selection of the best three performing ML algorithms in stage 1		Already covered with stage 1
SVR	X	X	X			
DT	X	X	X			
RF	X	X	X			
AB	X	X	X			
GB	X	X	X			
MLP	X	X	X			
Total No. of models	21			6		

regressor, an Adaptive Boosting (AB) regressor, a Gradient Boosting (GB) regressor and a Multilayer Perceptron (MLP). All of them are standard ML algorithms typically used in ML applications, but with a widespread across different learning techniques. RQ1 is primarily concerned with evaluating the feasibility in general of whether regression algorithms are suitable for predicting delivery delays. RQ2 focuses on the comparison of the performance of prediction models at different times of the prediction. To answer both RQs, it is therefore not necessary to maximize the performance of the ML model by sophisticated ML algorithms; simple and established ML algorithms are sufficient. Instead, it is necessary to ensure comparability across the three times the prediction. Thus, in all three considered times of prediction, the same set of ML algorithms is to be applied. Thus, in the first stage, to analyse the trade-off between the time of the prediction and the model quality, in total 21 prediction models were included in the experimental plan. Depending on the time of the prediction, the set of input features to be used for training and evaluation of the prediction model varied as defined in Table 6.2.

Further, the focus of the second stage of the experimental plan was to analyse the impact of a limitation in the scope of the prediction model in terms of a limitation in the product portfolio and supplier structure on the model quality. Here, we planned to compare the model quality of prediction models with different levels of limitations. In detail, we planned to compare the following three limitations:

- Limitation to *five or more* orders per component and supplier, following Brintrup et al. [20].
- Limitation to *two or more* orders per component and supplier.
- *No limitation*, which was applied in the 21 prediction models of the first step.

When limiting the product portfolio and supplier structure the number of samples varies, but the data structure itself remains equal. Further, in the second step, we were interested in the impact of the limitation and not in the impact of the ML algorithms nor the time of the prediction. Thus, we only planned to apply the limitations in scope to three selected prediction models of the first step of the experiment—the three best-performing models considering the time of the prediction and the model quality. Thus, in the second step, we included six more prediction models with different levels of limitation in the experimental plan— only six more models instead of nine since the level *No limitation* was already covered in the first step.

Consequently, in total, considering both stages of our experiment 27 prediction models were included in the experimental plan to quantify the impact of the time of the prediction, the impact of the ML algorithm, and the impact of a limitation in the product portfolio and supplier structure on the model quality. All prediction models were implemented in Python 3.7 utilizing the Scikit-learn library.

6.4.5   Evaluation and Results

In the evaluation phase, we trained and compared the performance of all prediction models following the two steps of our experimental plan. In both steps, we evaluated the reached model qualities utilizing MAE, RMSE, and R^2 as defined in the business understanding phase. Further, we split the data sets of all prediction models of the two steps into two separate train and test data sets with a ratio of 80 % for training and 20 % for testing the models utilizing the train_test_ split-function in sklearn with a fixed random_state. The train and test data sets were identical for each model to ensure comparability.

Subsequently, we trained the models of the first step based on the train data set and optimized the hyperparameters. For tuning the hyperparameters we used a grid search algorithm. An overview of the optimized hyperparameters used in each of the prediction models is given in the appendix in Table 6.6. After the training, we evaluated the achieved model qualities based on the test data set. The results of the first step are documented in Table 6.4. For the following, we defined a model as a trained ML algorithm at a specific time of the prediction. Additionally, for ease of reading, we used the simplified notation model$_{\text{ML algorithm, time of the prediction}}$. As an example, the model with the AB regressor as ML algorithm at the time of the prediction 'order placement' is notated as model$_{\text{AB, 1}}$.

Comparing the metrics revealed that with an R^2 of 92 % at the time of the prediction '1. purchasing request', and 98 % at the times of the prediction '2. order placement' and '3. delivery confirmation' the AB regressor performed best compared to other ML algorithms, closely followed by the RF regressor and GB regressor. Further, the performance of all ML algorithms increased from the time of the prediction '1. purchasing request' to '2. order placement' and remained almost equal from '2. order placement' to '3. delivery confirmation'. Thus, the information available after placing the order have an impact on the performance of the prediction model, whereas the information on the delivery confirmation has almost no impact. Consequently, the best performing model was model$_{\text{AB,2}}$ with information available after order placement. However, with an MAE of 8.2 days, an RSME of 19.9 days, and an R^2 of 92 % the model$_{\text{AB,1}}$ with information available after the purchasing request performed only slightly worse. Considering the trade-off between the time of the prediction and model quality, the possibility of taking actions earlier with a slightly lower model quality was preferable over a higher model quality in our case study. Thus, the model$_{\text{AB,1}}$ was our model of choice. Further, the reach model qualities already showed, that a prediction of delivery delays in calendar days using a regression approach is possible. Thus, with these results we can already answer the research questions RQ1 and RQ2:

- *Answer to RQ1:* Regression algorithms are capable to predict delivery delays of orders placed at suppliers of a low volume high variety manufacturer. Specifically, the prediction has an MAE of 8.2 to 3.2 days depending on the selected time of the prediction.
- *Answer to RQ2:* With a later time of the prediction and thus the available one of more information about the procurement process the prediction quality increases—which is trivial. Related to the MAE the prediction model improves by approx. 60 % (an improvement from 3.2 to 8.2 days) when comparing the creation of a purchasing request and the date of delivery confirmation by the supplier. Nevertheless, the model at the early time of prediction shows good prediction results.

For a practical application of the prediction model in the company under consideration, we suggest applying two models: First, a prediction after the purchasing request with a slightly lower model quality, followed by a second prediction after placing the order with a higher model quality. With this combination, the personnel in the purchasing department are supported with an early but less reliable risk assessment of potentially delayed deliveries which is updated after placing the order. Thus, we ensured both, a possibility to predict early and high model quality.

Table 6.4 Reached model qualities in the first step of the experiment

ML algorithm	Purchasing Request			Order Placement			Delivery Confirmation		
	MAE	RSME	R²	MAE	RSME	R²	MAE	RSME	R²
LR	29.5	43.7	59 %	19.0	27.8	83 %	19.3	28.2	83 %
SVR	29.7	43.9	58 %	19.1	27.8	83 %	19.3	28.3	82 %
DT	16.2	43.8	81 %	7.1	14.3	96 %	6.9	14.6	95 %
RF	10.2	20.6	91 %	4.6	10.7	98 %	4.2	11.0	98 %
AB	8.2	19.9	92 %	3.2	9.9	98 %	3.2	10.7	98 %
GB	10.9	20.1	91 %	4.8	9.9	98 %	4.6	10.6	98 %
MLP	25.2	35.9	72 %	9.4	14.8	95 %	9.2	15.0	95 %

Further, we analysed the feature importance of our model of choice (see Fig. 6.5) revealing those input features that significantly influence the model's output. The results showed that the features 'Order created' and 'BOM item created' are most important in the model$_{AB,1}$. At the time of the prediction—in this case, at the time of the creation of an internal purchasing request—the time of the demand creation of a component and the time of the transformation of the demand into a purchasing request are therefore decisive for the prediction of the delivery delay. Consequently, it can also be deduced that current delivery delays are largely determined by delays in demand creation and the transformation of the demand into a purchase order. For the practical application of our model at the machinery manufacturer, this indicated that the processes upstream of the purchasing process needed to be accelerated. Thus, based on the feature importance plot, we were able to identify the most relevant features for the prediction of delivery delays and to deduct general optimization potential at the machinery manufacturer in the processes upstream of the purchasing process.

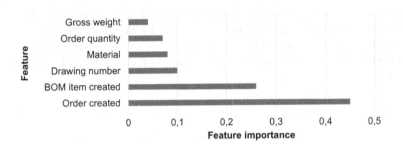

Fig. 6.5 Feature importance of the model$_{AB,1}$

After identifying model$_{AB,\,1}$ as the model of choice followed by model$_{RF,\,1}$ and model$_{GB,\,1}$, we continued with the second step of our experimental plan. Here, we applied the same training and evaluation procedure as in the first step—training and tuning the model based on the train data set and evaluating the model performance based on the test data set. Results are documented in Table 6.5. It was particularly noticeable that there is no increase nor decrease in the model quality when limiting the product portfolio or suppliers. Thus, in our case study, contrary to Brintrup et al. [20], a limitation was not necessary. Consequently, with these results we can answer RQ3 as follows:

- *Answer to RQ3:* In our exemplary case at the low volume high variety machinery manufacturer the curse of dimensionality was not an issue when setting up a regression model predicting delivery delays of orders placed at suppliers.

Subsequently, we further analysed our model of choice, model$_{AB,1}$, to gain a better understanding of its functionality. Here we conducted a sensitivity analysis of the dataset size (see Fig. 6.6) quantifying the relationship between dataset size and model performance. In detail, we varied the dataset size from 25 % to 100 % in 25 % steps of the 119,610 orders and evaluated the R^2-value. The results showed that the model quality was only slightly improving with a larger dataset size within the analysed range. Thus, a further increase in the dataset size might improve the model quality. Contrary, a reduction of the dataset size would also be acceptable since the model quality is reduced only minimally with a smaller dataset size. In our case study, the 119,610 orders were placed in three years. Thus, the manufacturer under consideration is placing approx. 40,000 orders per year. Since a dataset size of 25 % has provided almost equally good results, our model would also be applicable for a volume of 10,000 orders per year. Consequently, our model might also be applicable to manufacturers that place smaller numbers of orders.

Table 6.5   Reached model qualities in the second step of the experiment

ML algorithm	No limitation			Number of orders per component and supplier ≥ 2			Number of orders per component and supplier ≥ 5		
	MAE	RSME	R²	MAE	RSME	R²	MAE	RSME	R²
RF	10.2	20.6	91 %	10.1	21.1	90 %	10.1	21.1	90 %
AB	8.2	19.9	92 %	8.1	19.4	92 %	8.3	20.0	92 %
GB	10.9	20.1	91 %	10.9	20.4	91 %	10.2	19.8	91 %

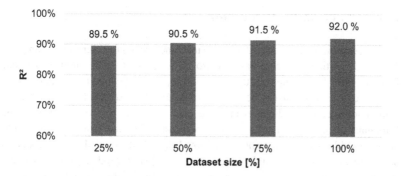

Fig. 6.6   Impact of the dataset size on the model quality

Overall, we were able to fulfil the objectives of our case study. In the first step, we confirmed that a regression model is capable to predict delivery delays and confirmed the influence of time of the prediction on the model performance. Further, we were able to select an AB regressor at the earliest time of the prediction 'purchasing request' as the model of choice, and thus, made a trade-off between the model quality and the time of the prediction. Thus, we could answer RQ1 with our first contribution that an ML-based regression model can predict delivery delays of orders placed at suppliers of machinery manufacturing in calendar days with an R^2 of 92 % in our model of choice. In addition, we could answer RQ2 with our second contribution that an early prediction within the purchasing process based on information available after creating the internal order request is possible and only slightly worse than a later prediction. Subsequently, we analysed the impact of a limitation of our data set on the model quality. Here, we could answer RQ3 with our third contribution that in our case study, there is no need to limit the dimensionality of our prediction model.

With our regression model, the company under consideration now can predict delivery delays immediately after creating a purchasing request and getting information about its severity. This helps to select adequate reactive measures in terms of costs for the measures and the predicted severities. In addition, companies can also predict the period a component will be delivered early. Thus, companies can allocate space for inventory and calculate days of inventory turnover predictably. Furthermore, since no restriction in dimensionality was necessary, the model is applicable to all components and suppliers considered.

6.5   Section IV: Discussion

In our research, we have considered one machinery manufacturer as an exemplary case study. Although a case-based approach is common in the field of machine learning (see, for example, [70, 71, 126, 128, 136]) the findings might remain case-specific and might not be generalizable to other sectors. Nevertheless, our study confirmed the findings of previous case studies of Brintrup et al. [20] and Baryannis et al. [21]—a prediction of potentially delayed deliveries of suppliers is possible. However, it should be noted that all three available case studies including ours focus on similar industries in terms of product complexity and quantity. In other industries, such as automotive, the product complexity and the number of components ordered may be different, potentially affecting the achievable model quality. Thus, further research should validate the achieved findings in other case studies and focus on different industries as well.

Further, we were able to prove that a limitation in the product portfolio and the supplier structure is not necessary. In our case, a limitation did not affect the model quality—neither positively nor negatively. Thus, further researchers can limit their scope if they want to reduce complexity but are not forced to do so by the curse of dimensionality if they use appropriate encoders for high-dimensional categorical features such as a binary encoder.

Moreover, our model quality with an R^2 of 92 % at the time of the prediction 'order request' and 98 % at the time 'order placement' is still afflicted with inaccuracy. Here, we identified the ML algorithm to be an important factor for model performance. Accordingly, the use of other regression ML algorithms could be a way to further improve the model quality. Future research should therefore investigate other regression ML algorithms for the prediction of potential delivery delays. Moreover, including additional information about the purchasing process in the ML model could further increase the model quality. For example, we considered so far only a little information about the required components—product number, weight, and material. Further information about the purchased components, such as their complexity or details from the drawing or CAD data, could improve the model performance. The trade-off between the effort in gathering additional data and improvements in the prediction model should also be further investigated by future research. In addition to data about the purchased component, information about the supplier's manufacturing process, such as the planned work schedule or available capacity, could add value to the prediction model as well. Future research could therefore try to set up an interface to the suppliers and thus include direct information from the supplier's manufacturing process in the prediction model to further increase its quality.

Further, the dataset of the case study originates from the ERP system of the considered machinery manufacturer. Here, the data fields of a purchasing order are overwritten with every update and there is no history of the recorded data. For example, a supplier can report an update of his expected delivery date, but the dataset in the system always contains only the last one reported. Consequently, it is possible that the dataset considered at the time of the prediction would not have been identical to the dataset that is available retrospectively. Therefore, future research should investigate whether overwriting the data set in ERP systems has an impact on the predictive performance of delivery delay prediction models. For this purpose, it is conceivable to record a data set with historical changes.

Moreover, a critical limitation for the practical applicability of the model is that the model in its current form only uses knowledge from the past to predict the future. However, if previously unknown events occur that have a significant impact on the supply chains (e.g., Covid, Ukraine war), the model will not take these influences into account and will make incorrect predictions. Therefore, with the current setting of the model, manual control is still necessary as a supplement to the model. Here, the integration of additional data sources such as the evaluation of the daily worldwide news would be beneficial.

6.6   Section V: Conclusion

Complex manufacturing industries that rely on externally sourced components need to ensure the on-time start of assembly processes, as delayed deliveries can cause costly assembly disruptions. To streamline operations that depend on external supply, the use of Machine Learning for the prediction of supply delays has been proposed in recent research with promising results. However, extant research considered classification approaches that depict whether a delay will occur but omitted its timing and duration. Furthermore, data-intensive approaches limit their application to high-volume settings whereas low volume high variety industries could also benefit from delay prediction. Finally, we posit that for a delay prediction to be meaningful, it needs to be early enough in the procurement cycle, such that mitigative actions can be taken.

In this work, we address these gaps through a systemic procedure setting up defined research questions and answering these with a two stages experimental plan containing a set of different ML regression models. Here, we show that regression algorithms are capable to predict delivery delays of suppliers of low volume high variety manufacturers. Additionally, we show that the severity of a delay can be predicted, early enough for facilitating action. Our approach also has the advantage of mitigating the curse of dimensionality, thereby making it applicable to low volume high variety settings.

For the development of the ML-based regression models, we followed the established Cross Industry Standard Process for Data Mining (CRISP-DM). Here, first, a detailed understanding of the company's procurement process was built and relevant features for our model were identified by performing a correlation analysis. Subsequently, we set up an extensive experimental plan to identify the best ML model. Our experimental plan consisted of two steps. The first step focused on the performance of different ML algorithms such as AB, RF and GB regressors at different times of the prediction to analyse whether a prediction in the early phases of the procurement process is possible. The second step included comparing different approaches to handle high-dimensional input features within a regression model. Executing the experimental plan revealed that an AB-regressor with an R^2 of 92 % trained on information available after the creation of an internal order request performed best, meaning that delay prediction can indeed be performed at the point of an order request.

Nevertheless, our model has scope for improvement. For example, the inclusion of further information such as component identifiers or supplier's manufacturing processes may further improve the prediction quality. The models have been tested on a single use case from the manufacturing industry. Further tests in low-volume-high-variety settings would increase confidence in the validity of our approach. Further, the model in its current form only uses knowledge from the past to predict the future. However, if previously unknown events occur that have a significant impact on the supply chains (e.g., Covid, Ukraine war), the model will not take these influences into account and will make incorrect predictions.

From a managerial standpoint, we recommend the use of regression models in the purchasing process, as it enables an early reaction to delivery delays even before an order is placed. Thus, companies can select adequate reactive measures in terms of cost and predicted severity. This can have a positive effect on the adherence to the start of assembly and ultimately on the adherence to the delivery date to the customer. In addition, companies can also predict the time a component will be delivered early. Thus, companies can allocate space for inventory and calculate days of inventory turnover predictably. Consequently, the regression model also can be used to reduce inventory costs.

6.7   Appendix

The hyperparameters used in the prediction models were optimized utilizing a grid search and cross-validation algorithms (GridSearchCV) from Scikit learn. Table 6.6 summarizes the utilized hyperparameters in the different prediction models.

Table 6.6 Hyperparameters of the prediction models

ML algorithm	Hyperparameter	Time of the prediction		
		1. Purchasing Request	2. Order Placement	3. Delivery Confirmation
SVR	C	1	1	1
	dual	True	True	True
	loss	squared epsilon insensitive	squared epsilon insensitive	squared epsilon insensitive
	Max iter	1000	1200	1200
DT	cc alpha	0.0028	0.0030	0.0032
	max features	auto	auto	auto
	max depth	25	25	25
	min samples split	4	4	4
	min samples leaf	5	5	5
RF	bootstrap	false	false	False
	max features	sqrt	sqrt	sqrt
	n estimators	200	220	230
AB	base estimator	decision tree (depth 3)	decision tree (depth 3)	decision tree (depth 3)
	n estimators	200	240	230
	learning rate	0.20	0.18	0.17
GB	n estimators	200	250	270
	max features	sqrt	sqrt	sqrt
	max depth	14	14	14

(continued)

Table 6.6 (continued)

ML algorithm	Hyperparameter	Time of the prediction		
		1. Purchasing Request	2. Order Placement	3. Delivery Confirmation
	min samples split	14	14	14
	min samples leaf	2	2	2
	learning rate	0.10	0.12	0.13
MLP	alpha	1,00E-06	1,00E-06	1,00E-06
	hidden layer size	35	37	34
	max iter	1000	1250	1150
	activation	relu	relu	relu
	solver	adam	adam	adam
	batch size	auto	Auto	auto
	learning rate	0.10	0.11	0.10

6.8   Presentation of Own Contribution

All the work presented within this publication was carried out by me. This includes the definition of research questions and hypotheses, the definition of the research approach including the set up to compare the performance of ML-models at different times of the prediction and different ML-algorithms, the collection, cleaning and aggregation of data from the manufacturing company under consideration, the feature engineering and setting up of the data model, the training and tuning of all ML-models including writing of the respective code in Python as well as the evaluation and critical reflection of the results. Further, transforming the work into text and writing of the publication were entirely in my hands. All co-authors Prof. Dr. Burggräf, Dr. Wagner, Mr. Heinbach, Mr. Saßmannshausen and Mrs. Brintrup contributed with ideas to the research concept and in the internal review process.

Critical Refection and Future Perspective

7

The four publications within the cumulative part of this thesis focus on the prediction of missing parts at the start of assembly. After the first systematic literature review in thesis I (cf. Chapter 3) to get a comprehensive overview of ML and OR approaches for lead time prediction, ML models for the prediction of missing parts at the start of assembly from in-house production and component procurement are presented in theses II—IV (cf. Chapters 4—6). However, the results within publications II-IV are based exclusively on case studies at two German machine manufacturers. In order to establish the validity of the findings on a broader basis, future work should include case studies of other machine manufacturers. In addition, it should be investigated whether the prediction of missing parts at the beginning of assembly is functional only for engineer-to-order manufacturers such as machine manufacturers or also for different types of manufacturing. Thus, an extension of the scope to other types of manufacturing, such as make-to-order, would be useful. It would be interesting to compare the necessary data fields, the model structures and the resulting model qualities for different types of manufacturers. Especially for serial manufacturers, the completion date of an individual component or the assembled product is not the primary focus. Instead, the quantity of products manufactured and the quality of the individual products play a central role. Therefore, in addition to models that predict schedule deviations, models that predict quantity and quality deviations would be helpful to ensure that customers are supplied on time, in the right quantity, and with the right quality.

In addition to the restriction on engineer-to-order manufacturers, only little information from the processes upstream of assembly are used within the three

© The Author(s), under exclusive license to Springer Fachmedien Wiesbaden GmbH, part of Springer Nature 2024
F. Steinberg, *Machine Learning-based Prediction of Missing Parts for Assembly*, Findings from Production Management Research ,
https://doi.org/10.1007/978-3-658-45033-5_7

case studies. This is particularly noticeable for information from the design process. Here, only information about the product name, the material or the weight was used within this cumulative promotion, which is information that is usually available in ERP or MES-systems. Since this little information already had a significant impact on the quality of the prediction models, it makes sense to add more information from the design process, such as dimensions, volumes, or the number and specification of features in the CAD model of the component, such as holes, shoulders, radii, or surface roughness.

Furthermore, this cumulative thesis has observed that classification approaches outperform regression approaches in predicting missing parts for in-house assembly. However, in addition to indicating whether a part is late, regression can also provide information about the severity of a delay. Therefore, it would be useful to develop approaches for predicting missing parts for in-house assembly that also use regression models to achieve better results. A starting point could be a detailed analysis of the reasons for the observed poorer performance of regression models compared to classification models in predicting missing parts for in-house assembly.

Summary

8

In this thesis, AI-based models were developed to predict missing parts from both the manufacturing and procurement processes. During the development of these models, the following scientific results were achieved in a total of four research papers:

First, a systematic literature review showed that the use of ML in lead time prediction is an area of research with increasing relevance. Order data and information about the status of the manufacturing system are mainly used. Information about the items to be produced—so-called material data—as well as feedback data from production are rarely used. Especially in complex forecasting models, which include several data sources as a broad spectrum of information, these data are not used yet.

Based on this, the influence of material data on the quality of models predicting missing parts at assembly start and the necessary level of detail of these models were investigated. Specifically, two case studies were conducted focusing on the material supply for the assembly of the in-house manufacturing process of a machine manufacturer. It was shown that a model with a low level of detail— i.e. less detailed models—using a classification approach leads to better results than more detailed models. It was also shown that material data has a significant impact on the quality of the model for predicting missing parts at the start of assembly.

Finally, a model for the prediction of defective parts in the procurement process was developed through a case study at a machine manufacturer. In contrast to existing approaches, an ML-based regression model for the prediction of delivery delays of orders placed with suppliers of the machine manufacturer in calendar days showed a satisfactory model quality. Furthermore, it was shown that an early

© The Author(s), under exclusive license to Springer Fachmedien Wiesbaden GmbH, part of Springer Nature 2024
F. Steinberg, *Machine Learning-based Prediction of Missing Parts for Assembly*,
Findings from Production Management Research ,
https://doi.org/10.1007/978-3-658-45033-5_8

prediction within the purchasing process based on information available after the creation of the internal order request is possible and only slightly worse than a later prediction.

Overall, this work developed the first machine learning model to predict missing parts at the start of assembly for in-house products. This model enables production controllers to identify delays in production orders at an early stage and to take measures to accelerate them. In addition, the first model predicting delivery delays using a regression approach was implemented as part of this work. By moving from a classification approach to a regression approach, it is now possible to predict the severity of a potential delay in addition to an indication that an order will be delivered late.

References

1. A. E. Coronado, A. C. Lyons, D. F. Kehoe, and J. Coleman, "Enabling mass customization: extending build-to-order concepts to supply chains," *Production Planning & Control*, vol. 15, no. 4, pp. 398–411, 2004, https://doi.org/10.1080/0953728042000238809.

2. P. Fredriksson and L.-E. Gadde, "Flexibility and rigidity in customization and build-to-order production," *Industrial Marketing Management*, vol. 34, no. 7, pp. 695–705, 2005, https://doi.org/10.1016/j.indmarman.2005.05.010.

3. P. Nyhuis, R. Nickel, and T. Busse, "Logistisches Controlling der Materialverfügbarkeit mit Bereitstellungsdiagrammen," *ZWF Zeitschrift für wirtschaftlichen Fabrikbetrieb*, no. 101, pp. 265–268, 2006.

4. H.-H. Wiendahl, *Auftragsmanagement der industriellen Produktion: Grundlagen, Konfiguration, Einführung*, 2011st ed. Berlin, Heidelberg: Springer Berlin Heidelberg, 2012.

5. G. Schuh and V. Stich, Eds., *Produktionsplanung und -steuerung*, 4th ed. Berlin, Heidelberg: Springer Vieweg, 2012.

6. A. Ripperger, *Gestaltung der Organisation effizienter Auftragsabwicklungsprozesse im Maschinen- und Anlagenbau unter typologischen Gesichtspunkten*. Aachen: Shaker, 1999.

7. G. Schuh and E. Westkämper, *Liefertreue im Maschinen- und Anlagenbau: Stand, Potenziale, Trends*. Stuttgart, 2006.

8. G. Reinhart, Ed., *Montage-Management: Lösungen zum Montieren am Standort Deutschland*. München: TCW, 1998.

9. M. Schmidt, *Modellierung logistischer Prozesse der Montage*. Zugl.: Hannover, Univ., Diss., 2010. Garbsen: PZH Produktionstechn. Zentrum, 2011.
pagebreak

10. H.-P. Wiendahl, *Betriebsorganisation für Ingenieure*. München: Carl Hanser Verlag GmbH & Co. KG, 2019.

11. P. Nyhuis and H.-P. Wiendahl, *Logistische Kennlinien: Grundlagen, Werkzeuge und Anwendungen*, 3rd ed. Berlin, Heidelberg: Springer, 2012.

12. H. Lödding, *Verfahren der Fertigungssteuerung: Grundlagen, Beschreibung, Konfiguration*, 2nd ed. Berlin, Heidelberg: Springer-Verlag, 2008.

F. Steinberg, *Machine Learning-based Prediction of Missing Parts for Assembly*, Findings from Production Management Research ,
https://doi.org/10.1007/978-3-658-45033-5

13. K. R. Baker and D. Trietsch, *Principles of sequencing and scheduling*. Hoboken, N.J: John Wiley, 2009.

14. Guilherme E. Vieira, Jeffrey W. Herrmann, and Edward Lin, "Rescheduling Manufacturing Systems: A Framework of Strategies, Policies, and Methods," *Journal of Scheduling*, vol. 6, no. 1, pp. 39–62, 2003, https://doi.org/10.1023/A:1022235519958.

15. A. Raddon and B. Grigsby, "Throughput time forecasting model," in *1997 IEEE/ SEMI Advanced Semiconductor Manufacturing Conference and Workshop ASMC 97 Proceedings*, Cambridge, MA, USA, Oct. 1997, pp. 430–433.

16. T. Berlec and M. Starbek, "Forecasting of Production Order Lead Time in Sme's," in *Products and Services*, I. Fuerstner, Ed., Rijeka: IntechOpen, 2010.

17. A. Pfeiffer, D. Gyulai, B. Kádár, and L. Monostori, "Manufacturing Lead Time Estimation with the Combination of Simulation and Statistical Learning Methods," *Procedia CIRP*, vol. 41, pp. 75–80, 2016, https://doi.org/10.1016/j.procir.2015.12.018.

18. S. M. Asadzadeh, A. Azadeh, and A. Ziaeifar, "A Neuro-Fuzzy-Regression Algorithm for Improved Prediction of Manufacturing Lead Time with Machine Breakdowns," *Concurrent Engineering*, vol. 19, no. 4, pp. 269–281, 2011, https://doi.org/10.1177/106 3293X11424512.

19. P. Burggräf, J. Wagner, B. Koke, and F. Steinberg, "Approaches for the prediction of lead times in an engineer to order environment – a systematic review," *IEEE Access*, vol. 8, pp. 142434–142445, 2020, https://doi.org/10.1109/ACCESS.2020.3010050.

20. A. Brintrup et al., "Supply chain data analytics for predicting supplier disruptions: a case study in complex asset manufacturing," *International Journal of Production Research*, vol. 58, no. 11, pp. 3330–3341, 2020, https://doi.org/10.1080/00207543. 2019.1685705.

21. G. Baryannis, S. Dani, and G. Antoniou, "Predicting supply chain risks using machine learning: The trade-off between performance and interpretability," *Future Generation Computer Systems*, vol. 101, pp. 993–1004, 2019, https://doi.org/10.1016/j.future. 2019.07.059.

22. R. Carnap, "Formalwissenschaft und Realwissenschaft," *Erkenntnis*, vol. 5, no. 1, pp. 30–37, 1935, https://doi.org/10.1007/BF00172279.

23. P. Ulrich and W. Hill, "Wissenschaftstheoretische Grundlagen der Betriebswirtschaftslehre," *Wirtschaftswissenschaftliches Studium: Zeitschrift für Ausbildung und Hochschulkontakt*, vol. 5, 7+8, pp. 304–309, 1976.

24. H. A. Simon, *The sciences of the artificial*, 3rd ed.: MIT Press, 1996.

25. J. Vom Brocke, A. Hevner, and A. Maedche, *Design Science Research. Cases*. Cham: Springer International Publishing, 2020.

26. Hevner, March, Park, and Ram, "Design Science in Information Systems Research," *MIS Quarterly*, vol. 28, no. 1, p. 75, 2004, https://doi.org/10.2307/25148625.

27. M. Rudberg and J. Wikner, "Mass customization in terms of the customer order decoupling point," *Production Planning & Control*, vol. 15, no. 4, pp. 445–458, 2004, https:// doi.org/10.1080/0953728042000238764.

28. H.-P. Wiendahl, *Load-Oriented Manufacturing Control*. Berlin, Heidelberg, s.l.: Springer Berlin Heidelberg, 1995.

29. G. Schuh and V. Stich, *Logistikmanagement: Handbuch Produktion und Management 6*, 2nd ed. Berlin, Heidelberg: Springer, 2013.

30. W. Eversheim and G. Schuh, *Produktion und Management Betriebshütte*, 7th ed. Berlin, Heidelberg: Springer, 1996.

31. V. Stich, J. Quick, and S. Cuber, "Konfiguration logistischer Netzwerke," in *Logistik-management: Handbuch Produktion und Management 6*, G. Schuh and V. Stich, Eds., Berlin, Heidelberg: Springer, 2013, pp. 35–76.

32. S. Beck, *Modellgestütztes Logistikcontrolling konvergierender Materialflüsse*. Zugl.: Hannover, Univ., Diss., 2013. Garbsen: PZH-Verl., 2013.

33. J. Feldhusen and K.-H. Grote, *Pahl/Beitz Konstruktionslehre: Methoden und Anwendung erfolgreicher Produktentwicklung*, 8th ed. Berlin, Heidelberg: Springer, 2013.

34. J. O. Fischer, *Kostenbewusstes Konstruieren: Praxisbewährte Methoden und Informationssysteme für den Konstruktionsprozess*, 1st ed. Berlin, Heidelberg: Springer, 2008.

35. G. Pahl, W. Beitz, J. Feldhusen, and K.-H. Grote, *Konstruktionslehre: Grundlagen erfolgreicher Produktentwicklung. Methoden und Anwendung*, 7th ed. Berlin [u. a.]: Springer, 2006.

36. W. Eversheim, *Organisation in der Produktionstechnik*, 3rd ed. Düsseldorf: VDI-Verl., 1996.

37. D. Arnold, K. Furmans, H. Isermann, and A. Kuhn, *Handbuch Logistik*, 3rd ed. Berlin, Heidelberg: Springer-Verlag, 2008.

38. P. Nyhuis, H.-P. Wiendahl, T. Fiege, and H. Mühlenbruch, "Materialbereitstellung in der Montage," in *Montage in der industriellen Produktion*, B. Lotter and H.-P. Wiendahl, Eds., Berlin, Heidelberg: Springer Berlin Heidelberg, 2006, pp. 323–351.

39. C. Frühwald and C. Wolter, "Prozessgestaltung," in *Prozessmanagement in der Wertschöpfungskette*, N. Hagen, P. Nyhuis, C. Frühwald, and M. Felder, Eds., 1st ed., Bern: Haupt, 2006, pp. 51–78.

40. P. Nyhuis and H. Rottbauer, "Erfolgsfaktoren und Hebel der Beschaffung im Rahmen eines Integrated Supply Managements," in *Integrated Supply Management: Einkauf und Beschaffung: Effizienz steigern, Kosten senken*, R. Bogaschewsky, Ed., Köln: Dt. Wirtschaftsdienst, 2003, pp. 117–138.

41. H.-J. Warnecke, *Der Produktionsbetrieb*. Berlin: Springer, 1995.

42. D. E. Whitney, *Mechanical assemblies: Their design, manufacture, and role in product development*. New York: Oxford Univ. Press, 2004.

43. REFA Verband für Arbeitsgestaltung, Betriebsorganisation und Unternehmensentwicklung e. V., *Methodenlehre der Planung und Steuerung: Teil 2*. München: Hanser, 1991.

44. *Montage- und Handhabungstechnik; Handhabungsfunktionen, Handhabungseinrichtungen; Begriffe, Definitionen, Symbole*, 2860, VDI, May. 1990.

45. J. H. Blackstone, *APICS Dictionary*, 14th ed. Chicago, IL: APICS, 2013.

46. G. Reinhart, R. Cuiper, and M. Loferer, "Die Bedeutung der Montage als letztes Glied in der Auftragsabwicklung," in *TCW-Report*, vol. 6, *Montage-Management: Lösungen zum Montieren am Standort Deutschland*, G. Reinhart, Ed., München: TCW, 1998, pp. 7–11.

47. J. Benz and M. Höflinger, *Logistikprozesse mit SAP®: Eine anwendungsbezogene Einführung; mit durchgehendem Fallbeispiel; geeignet für SAP Version 4.6A bis ECC 6.0*, 3rd ed. Wiesbaden: Vieweg+Teubner, 2011.

48. H. Kettner, *Neue Wege der Bestandsanalyse im Fertigungsbereich*, 1976.

49. G. W. Plossl, "Manufacturing Control," *The Last Frontier for Profits. Reston, VA: Reston Publishing*, 1973.

50. H.-H. Wiendahl, *Situative Konfiguration des Auftragsmanagements im turbulenten Umfeld*. Heimsheim: Jost-Jetter, 2002.

51. R. Nickel, *Logistische Modelle für die Montage*. Garbsen: PZH Produktionstechn. Zentrum, 2008.

52. H. Lödding, *Handbook of manufacturing control: Fundamentals, description, configuration*. Heidelberg: Springer, 2013.

53. O. Kocatepe, "An application framework for scheduling optimization problems," in *AICT: 2014 IEEE 8th International Conference on Application of Information and Communication Technologies: 15–17 October 2014*, Astana, Kazakhstan, 2015, pp. 1–4.

54. A. Schömig, D. Eichhorn, and G. Obermaier, "Über verschiedene Ansätze zur Ermittlung von Betriebskennlinien – Eine Anwendungsstudie aus der Halbleiterindustrie," in *Operations Research Proceedings*, v.2006, *: Selected Papers of the Annual International Conference of the German Operations Research Society (GOR), Jointly Organized with the Austrian Society of Operations Research (ÖGOR) and the Swiss Society of Operatio*, K.-H. Waldmann and U. M. Stocker, Eds., 1st ed., s.l.: Springer-Verlag, 2007, pp. 467–472.

55. C. Engelhardt, *Betriebskennlinien: Produktivität steigern in der Fertigung*. München: Hanser, 2000.

56. S. S. Aurand and P. J. Miller, "The operating curve: a method to measure and benchmark manufacturing line productivity," in *1997 IEEE/SEMI Advanced Semiconductor Manufacturing Conference and Workshop ASMC 97 Proceedings*, Cambridge, MA, USA, Oct. 1997, pp. 391–397.

57. J. Fowler and J. Robinson, "Measurement and improvement of manufacturing capacities (MIMAC): Final report," Technical Report 95062861A-TR, SEMATECH, Austin, TX, 1995.

58. C. Reuter and F. Brambring, "Improving Data Consistency in Production Control," *Procedia CIRP*, vol. 41, pp. 51–56, 2016, https://doi.org/10.1016/j.procir.2015.12.116.

59. S. Lee, Y. J. Kim, T. Cheong, and S. H. Yoo, "Effects of Yield and Lead-Time Uncertainty on Retailer-Managed and Vendor-Managed Inventory Management," *IEEE Access*, vol. 7, pp. 176051–176064, 2019, https://doi.org/10.1109/ACCESS.2019.295 7595.

60. W. Raaymakers and A. Weijters, "Makespan estimation in batch process industries: A comparison between regression analysis and neural networks," *European Journal of Operational Research*, vol. 145, no. 1, pp. 14–30, 2003, https://doi.org/10.1016/S0377-2217(02)00173-X.

61. A. Kampker, J. Wagner, P. Burggräf, and Y. Bäumers, "Criticality-focused, pre-emptive disruption management in low-volume assembly," *Proceedings of Abstract and Papers of 23rd International Conference on Production Research ICPR23 – Operational Excellence towards sustainable development goals (SDG) through Industry 4.0, Manila/Philippines*, no. 4, pp. 2–5, 2015.

62. G. Schuh, T. Potente, and T. Jasinski, "Decentralized, Market-Driven coordination mechanism based on the monetary value of in time deliveries," *Proceedings of Global Business Research, Kathmandu*, pp. 1–13, 2013.

63. R. J. Abumaizar and J. A. Svestka, "Rescheduling job shops under random disruptions," *International Journal of Production Research*, vol. 35, no. 7, pp. 2065–2082, 1997, https://doi.org/10.1080/002075497195074.

64. P. Burggräf, J. Wagner, K. Lück, and T. Adlon, "Cost-benefit analysis for disruption prevention in low-volume assembly," *Production Engineering Research and Development*, vol. 11, no. 3, pp. 331–342, 2017, https://doi.org/10.1007/s11740-017-0735-6.

65. J. Wagner, P. Burggräf, M. Dannapfel, and C. Fölling, "Assembly Disruptions: Empirical Evidence in the Manufacturing Industry of Germany, Austria and Switzerland," *International Refereed Journal of Engineering and Science: IRJES*, vol. 6, pp. 15–25, 2017.

66. N. Levin and J. Zahavi, "Predictive modeling using segmentation," *Journal of Interactive Marketing*, vol. 15, no. 2, pp. 2–22, 2001, https://doi.org/10.1002/dir.1007.

67. I. Goodfellow, Y. Bengio, and A. Courville, *Deep learning*. Cambridge, Massachusetts, London, England: MIT Press, 2016.

68. T. Cheng and M. C. Gupta, "Survey of scheduling research involving due date determination decisions," *European Journal of Operational Research*, vol. 38, no. 2, pp. 156–166, 1989, https://doi.org/10.1016/0377-2217(89)90100-8.

69. A. Öztürk, S. Kayalıgil, and N. E. Özdemirel, "Manufacturing lead time estimation using data mining," *European Journal of Operational Research*, vol. 173, no. 2, pp. 683–700, 2006, https://doi.org/10.1016/j.ejor.2005.03.015.

70. L. Lingitz *et al.*, "Lead time prediction using machine learning algorithms: A case study by a semiconductor manufacturer," *Procedia CIRP*, vol. 72, pp. 1051–1056, 2018, https://doi.org/10.1016/j.procir.2018.03.148.

71. A. D. Karaoglan and O. Karademir, "Flow time and product cost estimation by using an artificial neural network (ANN): A case study for transformer orders," *The Engineering Economist*, vol. 62, no. 3, pp. 272–292, 2017, https://doi.org/10.1080/0013791X.2016.1185808.

72. J. Vom Brocke *et al.*, "Reconstructing the giant: On the importance of rigour in documenting the literature search process," *ECIS 2009 Proceedings*, 2009.

73. D. Moher, A. Liberati, J. Tetzlaff, and D. G. Altman, "Preferred reporting items for systematic reviews and meta-analyses: The PRISMA statement," *International Journal of Surgery*, vol. 8, no. 5, pp. 336–341, 2010, https://doi.org/10.1016/j.ijsu.2010.02.007.

74. T. Weißer, T. Saßmannshausen, D. Ohrndorf, P. Burggräf, and J. Wagner, "A clustering approach for topic filtering within systematic literature reviews," *MethodX*, vol. 7, no. 100831, 2020, https://doi.org/10.1016/j.mex.2020.100831.

75. Business Dictionary. [Online]. Available: http://www.businessdictionary.com/ (accessed: May 15 2020).

76. S. O'Shea, Ed., *Cambridge business english dictionary*. Stuttgart, Cambrigde: Ernst Klett Sprachen GmbH; Cambridge Unversity Press, 2011.

77. A. Gunasekaran, C. Patel, and E. Tirtiroglu, "Performance measures and metrics in a supply chain environment," *International Journal of Operations & Production Management*, vol. 21, 1/2, pp. 71–87, 2001, https://doi.org/10.1108/01443570110358468.

78. S. M. Meerkov and C.-B. Yan, "Production Lead Time in Serial Lines: Evaluation, Analysis, and Control," *IEEE Trans. Automat. Sci. Eng.*, vol. 13, no. 2, pp. 663–675, 2016, https://doi.org/10.1109/TASE.2014.2365108.

79. G. Schuh, J.-P. Prote, F. Sauermann, and B. Franzkoch, "Databased prediction of order-specific transition times," *CIRP Annals*, vol. 68, no. 1, pp. 467–470, 2019, https://doi.org/10.1016/j.cirp.2019.03.008.

80. W. Bechte, *Steuerung der Durchlaufzeit durch belastungsorientierte Auftragsfreigabe bei Werkstattfertigung:(Rückentitel: Belastungsorientierte Auftragsfreigabe)*: VDI-Verlag, 1984.

81. J. F. C. Kingman, "The single server queue in heavy traffic," *Math. Proc. Camb. Phil. Soc.*, vol. 57, no. 4, pp. 902–904, 1961, https://doi.org/10.1017/S0305004100036094.

82. W. J. Hopp and M. L. Spearman, *Factory physics,* 3rd ed. Long Grove, Ill: Waveland Press, 2011.

83. L. Sattler, "Using queueing curve approximations in a fab to determine productivity improvements," in *1996 IEEE/Semi Advanced Semiconductor Manufacturing Conference and Workshop*, Cambridge, MA, USA, Nov. 1996, pp. 140–145.

84. F. G. Boebel and O. Ruelle, "Cycle time reduction program at ACL," in pp. 165–168.

85. D. W. Collins, K. Williams, and F. C. Hoppensteadt, "Implementation of Minimum Inventory Variability Scheduling 1-Step Ahead Policy(R) in a large semiconductor manufacturing facility," in *1997 IEEE 6th International Conference on Emerging Technologies and Factory Automation proceedings, EFTA '97, UCLA Conference Center, Los Angeles, California, September 9–12, 1997*, Los Angeles, CA, USA, 1997, pp. 497–504.

86. O. Ruelle, "Continuous flow manufacturing: the ultimate theory of constraints," in *1997 IEEE/SEMI Advanced Semiconductor Manufacturing Conference and Workshop ASMC 97 Proceedings*, Cambridge, MA, USA, Oct. 1997, pp. 216–221.

87. O. Rose, "The shortest processing time first (SPTF) dispatch rule and some variants in semiconductor manufacturing," in *Proceedings of the 2001 Winter Simulation Conference: Crystal Gateway Marriott, Arlington, VA, U.S.A., 9–12 December, 2001*, Arlington, VA, USA, 2001, pp. 1220–1224.

88. K. Wu and L. McGinnis, "Performance evaluation for general queueing networks in manufacturing systems: Characterizing the trade-off between queue time and utilization," *European Journal of Operational Research*, vol. 221, no. 2, pp. 328–339, 2012, https://doi.org/10.1016/j.ejor.2012.03.019.

89. J. V. Leon, D. S. Wu, and R. H. Storer, "Robustness measures and robust scheduling for job shops," *IIE Transactions*, vol. 26, no. 5, pp. 32–43, 1994, https://doi.org/10.1080/07408179408966626.

90. S. Tadayonirad, H. Seidgar, H. Fazlollahtabar, and R. Shafaei, "Robust scheduling in two-stage assembly flow shop problem with random machine breakdowns: integrated meta-heuristic algorithms and simulation approach," *AA*, vol. 39, no. 5, pp. 944–962, 2019, https://doi.org/10.1108/AA-10-2018-0165.

91. R. Conway, "Priority dispatching and job lateness in a job shop," *Journal of Industrial Engineering*, no. 16, pp. 228–237, 1965.

92. J. K. Weeks, "A Simulation Study of Predictable Due-Dates," *Management Science*, vol. 25, no. 4, pp. 363–373, 1979, https://doi.org/10.1287/mnsc.25.4.363.

93. J. Bertrand, "The use of workload information to control job lateness in controlled and uncontrolled release production systems," *Journal of Operations Management*, vol. 3, no. 2, pp. 79–92, 1983, https://doi.org/10.1016/0272-6963(83)90009-8.

94. G. L. Ragatz and V. A. Mabert, "A simulation analysis of due date assignment rules," *Journal of Operations Management*, vol. 5, no. 1, pp. 27–39, 1984, https://doi.org/10.1016/0272-6963(84)90005-6.

95. S. Eilon and I. G. Chowdhury, "Due dates in job shop scheduling," *International Journal of Production Research*, vol. 14, no. 2, pp. 223–237, 1976, https://doi.org/10.1080/00207547608956596.

96. John D. C. Little, "A Proof for the Queuing Formula: L= λ W," *Operations Research*, vol. 9, no. 3, pp. 383–387, 1961.

97. P. Burggräf, J. Wagner, and B. Koke, "Artificial intelligence in production management: A review of the current state of affairs and research trends in academia," in *2018 International Conference on Information Management and Processing (ICIMP)*, London, 2018, pp. 82–88.

98. A. Alenezi, S. A. Moses, and T. B. Trafalis, "Real-time prediction of order flowtimes using support vector regression," *Computers & Operations Research*, vol. 35, no. 11, pp. 3489–3503, 2008, https://doi.org/10.1016/j.cor.2007.01.026.

99. C. Wang and P. Jiang, "Deep neural networks based order completion time prediction by using real-time job shop RFID data," *J Intell Manuf*, vol. 30, no. 3, pp. 1303–1318, 2019, https://doi.org/10.1007/s10845-017-1325-3.

100. R. Armstrong, B. J. Hall, J. Doyle, and E. Waters, "Cochrane Update. 'Scoping the scope' of a cochrane review," *Journal of public health (Oxford, England)*, vol. 33, no. 1, pp. 147–150, 2011, https://doi.org/10.1093/pubmed/fdr015.

101. J. Higgins, J. Thomas, and J. Chandler, *Cochrane handbook for systematic reviews of interventions*. Chichester: John Wiley & Sons, 2019.

102. A. P. Siddaway, A. M. Wood, and L. V. Hedges, "How to Do a Systematic Review: A Best Practice Guide for Conducting and Reporting Narrative Reviews, Meta-Analyses, and Meta-Syntheses," *Annual review of psychology*, vol. 70, pp. 747–770, 2019, https://doi.org/10.1146/annurev-psych-010418-102803.

103. G. Paré, M.-C. Trudel, M. Jaana, and S. Kitsiou, "Synthesizing information systems knowledge: A typology of literature reviews," *Information & Management*, vol. 52, no. 2, pp. 183–199, 2015, https://doi.org/10.1016/j.im.2014.08.008.

104. H. M. Cooper, "Organizing knowledge syntheses: A taxonomy of literature reviews," *Knowledge in Society*, vol. 1, no. 1, pp. 104–126, 1988, https://doi.org/10.1007/BF03177550.

105. R. J. Torraco, "Writing Integrative Literature Reviews: Guidelines and Examples," *Human Resource Development Review*, vol. 4, no. 3, pp. 356–367, 2005, https://doi.org/10.1177/1534484305278283.

106. J. Rowley and F. Slack, "Conducting a literature review," *Management Research News*, vol. 27, no. 6, pp. 31–39, 2004, https://doi.org/10.1108/01409170410784185.

107. A. Booth, ""Brimful of STARLITE": toward standards for reporting literature searches," *Journal of the Medical Library Association*, vol. 94, no. 4, 421-e205, 2006.

108. C. C. Aggarwal and C. Zhai, "A Survey of Text Clustering Algorithms," in *Mining Text Data*, C. C. Aggarwal and C. Zhai, Eds., 2012nd ed., Boston, MA: Springer US, 2012, pp. 77–128.

109. S. Adinugroho, Y. A. Sari, M. A. Fauzi, and P. P. Adikara, "Optimizing K-means text document clustering using latent semantic indexing and pillar algorithm," in *2017 5th*

International Symposium on Computational and Business Intelligence (ISCBI), Dubai, United Arab Emirates, Aug. 2017–Aug. 2017, pp. 81–85.

110. Y. Levy and T. J. Ellis, "A Systems Approach to Conduct an Effective Literature Review in Support of Information Systems Research," *InformingSciJ*, vol. 9, pp. 181–212, 2006, https://doi.org/10.28945/479.

111. P. Salipante, W. Notz, and J. Bigelow, "A matrix approach to literature reviews," *Research in organizational behavior: an annual series of analytical essays and critical reviews*, vol. 4, 1982.

112. J. Webster and R. T. Watson, "Analyzing the Past to Prepare for the Future: Writing a Literature Review," *MIS Quarterly*, vol. 26, no. 2, pp. xiii–xxiii, 2002.

113. L. Cronjäger, Ed., *Bausteine für die Fabrik der Zukunft: Eine Einführung in die rechnerintegrierte Produktion (CIM)*. Berlin, Heidelberg: Springer Berlin Heidelberg, 1994.

114. G. Ioannou and S. Dimitriou, "Lead time estimation in MRP/ERP for make-to-order manufacturing systems," *International Journal of Production Economics*, vol. 139, no. 2, pp. 551–563, 2012, https://doi.org/10.1016/j.ijpe.2012.05.029.

115. H. Sabeti and F. Yang, "Flow-time estimation by synergistically modeling real and simulation data," in *2017 Winter Simulation Conference (WSC)*, Las Vegas, NV, Dec. 2017 - Dec. 2017, pp. 3230–3241.

116. N. Govind and T. Roeder, "Estimating Expected Completion Times with Probabilistic Job Routing," in *Proceedings of the 2006 Winter Simulation Conference*, Monterey, CA, USA, Dec. 2016, pp. 1804–1810.

117. D. H. Grabenstetter and J. M. Usher, "Determining job complexity in an engineer to order environment for due date estimation using a proposed framework," *International Journal of Production Research*, vol. 51, no. 19, pp. 5728–5740, 2013, https://doi.org/10.1080/00207543.2013.787169.

118. W. Zimmermann and U. Stache, *Operations-Research: Quantitative Methoden zur Entscheidungsvorbereitung*, 10th ed. München: Oldenbourg, 2001.

119. G. Feichtinger and R. F. Hartl, *Optimale Kontrolle ökonomischer Prozesse: Anwendungen des Maximumprinzips in den Wirtschaftswissenschaften*. Berlin: W. de Gruyter, 1986.

120. R. Caruana and A. Niculescu-Mizil, "An empirical comparison of supervised learning algorithms," in *Proceedings of the 23rd international conference on Machine learning – ICML '06*, Pittsburgh, Pennsylvania, 2006, pp. 161–168.

121. V. Jain and T. Raj, "An adaptive neuro-fuzzy inference system for makespan estimation of flexible manufacturing system assembly shop: a case study," *Int J Syst Assur Eng Manag*, vol. 9, no. 6, pp. 1302–1314, 2018, https://doi.org/10.1007/s13198-018-0729-6.

122. T. Berlec, E. Govekar, J. Grum, P. Potočnik, and M. Starbek, "Predicting Order Lead Times," *Journal of Mechanical Engineering*, vol. 54, no. 5, pp. 308–321, 2008.

123. J. Gramdi, "Elaborating actual lead time with both management and execution data," in *2009 International Conference on Computers & Industrial Engineering*, Troyes, France, Jun. 2009, pp. 674–677.

124. D. Gyulai, A. Pfeiffer, G. Nick, V. Gallina, W. Sihn, and L. Monostori, "Lead time prediction in a flow-shop environment with analytical and machine learning approaches,"

IFAC-PapersOnLine, vol. 51, no. 11, pp. 1029–1034, 2018, https://doi.org/10.1016/j.ifacol.2018.08.472.

125. W. Weng and S. Fujimura, "Estimating Job Flow Times by Using an Agent-Based Approach," in *2016 5th IIAI International Congress on Advanced Applied Informatics (IIAI-AAI)*, Kumamoto, Japan, Jun. 2016, pp. 975–979.

126. S. Singh and U. Soni, "Predicting Order Lead Time for Just in Time production system using various Machine Learning Algorithms: A Case Study," in *2019 9th International Conference on Cloud Computing, Data Science & Engineering (Confluence)*, Noida, India, Jan. 2019, pp. 422–425.

127. W. Fang, Y. Guo, W. Liao, K. Ramani, and S. Huang, "Big data driven jobs remaining time prediction in discrete manufacturing system: a deep learning-based approach," *International Journal of Production Research*, vol. 13, no. 1, pp. 1–16, 2019, https://doi.org/10.1080/00207543.2019.1602744.

128. I. Tirkel, "Forecasting flow time in semiconductor manufacturing using knowledge discovery in databases," *International Journal of Production Research*, vol. 51, no. 18, pp. 5536–5548, 2013, https://doi.org/10.1080/00207543.2013.787168.

129. I. Tirkel, "Cycle time prediction in wafer fabrication line by applying data mining methods," in *IEEE/SEMI Advanced Semiconductor Manufacturing Conference*, Saratoga Springs, NY, USA, May. 2011 - May. 2011, pp. 1–5.

130. A. Aburomman, M. Lama, and A. Bugarin, "A Vector-Based Classification Approach for Remaining Time Prediction in Business Processes," *IEEE Access*, vol. 7, pp. 128198–128212, 2019, https://doi.org/10.1109/ACCESS.2019.2939631.

131. T. Berlec and M. Starbek, "Predicting Order Due Date," *Arab J Sci Eng*, vol. 37, no. 6, pp. 1751–1766, 2012, https://doi.org/10.1007/s13369-012-0279-1.

132. F. Steinberg, P. Burggaef, J. Wagner, and B. Heinbach, "Impact of material data in assembly delay prediction—a machine learning-based case study in machinery industry," *Int J Adv Manuf Technol*, vol. 120, 1–2, pp. 1333–1346, 2022, https://doi.org/10.1007/s00170-022-08767-3.

133. D. X. Peng and G. Lu, "Exploring the Impact of Delivery Performance on Customer Transaction Volume and Unit Price: Evidence from an Assembly Manufacturing Supply Chain," *Prod Oper Manag*, vol. 26, no. 5, pp. 880–902, 2017, https://doi.org/10.1111/poms.12682.

134. A. T. Joseph, "Formulation of Manufacturing Strategy," *Int J Adv Manuf Technol*, vol. 15, no. 7, pp. 522–535, 1999, https://doi.org/10.1007/s001700050098.

135. S. Y. Nof, W. E. Wilhelm, and H.-J. Warnecke, *Industrial Assembly*. Boston, MA, s.l.: Springer US, 1997.

136. P. Burggräf, J. Wagner, B. Heinbach, and F. Steinberg, "Machine Learning-Based Prediction of Missing Components for Assembly – a Case Study at an Engineer-to-Order Manufacturer," *IEEE Access*, vol. 9, pp. 105926–105938, 2021, https://doi.org/10.1109/ACCESS.2021.3075620.

137. A. Zheng and A. Casari, *Feature engineering for machine learning: Principles and techniques for data scientists*. Beijing, Boston: O'Reilly, 2018.

138. M. F. Kilkenny and K. M. Robinson, "Data quality: "Garbage in – garbage out"," *Health information management: journal of the Health Information Management Association of Australia*, vol. 47, no. 3, pp. 103–105, 2018, https://doi.org/10.1177/1833358318774357.

139. J. Gosling and M. M. Naim, "Engineer-to-order supply chain management: A literature review and research agenda," *International Journal of Production Economics*, vol. 122, no. 2, pp. 741–754, 2009, https://doi.org/10.1016/j.ijpe.2009.07.002.

140. N. Slack, A. Brandon-Jones, and R. Johnston, *Operations management*, 8th ed. Harlow, England, London, New York: Pearson, 2016.

141. Y. Sun, C. Zhang, L. Gao, and X. Wang, "Multi-objective optimization algorithms for flow shop scheduling problem: a review and prospects," *The International Journal of Advanced Manufacturing Technology*, vol. 55, 5–8, pp. 723–739, 2011, https://doi.org/10.1007/s00170-010-3094-4.

142. W. Wang, G. Tian, G. Yuan, and D. T. Pham, "Energy-time tradeoffs for remanufacturing system scheduling using an invasive weed optimization algorithm," *Journal of Intelligent Manufacturing*, vol. 137, no. 3, p. 1602, 2021, https://doi.org/10.1007/s10845-021-01837-5.

143. G. Tian *et al.*, "Operation patterns analysis of automotive components remanufacturing industry development in China," *Journal of Cleaner Production*, vol. 164, no. 3, pp. 1363–1375, 2017, https://doi.org/10.1016/j.jclepro.2017.07.028.

144. D. Y. Sha and C.-H. Liu, "Using Data Mining for Due Date Assignment in a Dynamic Job Shop Environment," *The International Journal of Advanced Manufacturing Technology*, vol. 25, 11–12, pp. 1164–1174, 2005, https://doi.org/10.1007/s00170-003-1937-y.

145. D. M. McCutcheon and J. R. Meredith, "Conducting case study research in operations management," *Journal of Operations Management*, vol. 11, no. 3, pp. 239–256, 1993, https://doi.org/10.1016/0272-6963(93)90002-7.

146. C. Shearer, "The CRISP-DM model: the new blueprint for data mining," *Journal of data warehousing*, vol. 5, no. 4, pp. 13–22, 2000.

147. R. Wirth and J. Hipp, "CRISP-DM: Towards a standard process model for data mining," *Proceedings of the 4th international conference on the practical applications of knowledge discovery and data mining*, no. 1, 2000.

148. J. Davis and M. Goadrich, "The relationship between Precision-Recall and ROC curves," in *Proceedings of the 23rd international conference on Machine learning – ICML '06*, Pittsburgh, Pennsylvania, 2006, pp. 233–240.

149. D. M. W. Powers, "Evaluation: from precision, recall and F-measure to ROC, informedness, markedness and correlation," *International Journal of Machine Learning Technology 2:1 (2011)*, pp. 37–63, 2020. [Online]. Available: arXiv-ID 2010.16061

150. D. Chicco and G. Jurman, "The advantages of the Matthews correlation coefficient (MCC) over F1 score and accuracy in binary classification evaluation," *BMC genomics*, vol. 21, no. 1, p. 6, 2020, https://doi.org/10.1186/s12864-019-6413-7.

151. M. Kuhn and K. Johnson, *Applied predictive modeling*, 5th ed. New York: Springer, 2016.

152. J. Han, M. Kamber, and J. Pei, *Data mining: Concepts and techniques*, 3rd ed. Amsterdam: Elsevier/Morgan Kaufmann, 2012.

153. G. Dong and H. Liu, *Feature Engineering for Machine Learning and Data Analytics*. Milton: Chapman and Hall/CRC, 2018.

154. V. Verdhan, *Supervised Learning with Python: Concepts and Practical Implementation Using Python*, 1st ed. Berkeley CA: Apress; Imprint: Apress, 2020.

155. M. M. Mukaka, "A guide to appropriate use of Correlation coefficient in medical research," *Malawi Medical Journal: The Journal of Medical Association of Malawi,* vol. 24, no. 3, pp. 69–71, 2012.
156. D. Hinkle, S. G. Jurs, and W. Wiersma, *Applied statistics for behavioural sciences,* 5th ed. Boston: Houghton Mifflin Company, 2003.
157. J. Mohamad-Saleh and B. S. Hoyle, "Improved neural network performance using principal component analysis on Matlab," *International journal of the computer, the internet and Management,* vol. 16, no. 2, pp. 1–8, 2008.
158. I. Basheer and M. Hajmeer, "Artificial neural networks: fundamentals, computing, design, and application," *Journal of Microbiological Methods,* vol. 43, no. 1, pp. 3–31, 2000, https://doi.org/10.1016/S0167-7012(00)00201-3.
159. P. Jeatrakul and K. W. Wong, "Comparing the performance of different neural networks for binary classification problems," in *2009 Eighth International Symposium on Natural Language Processing,* Bangkok, Thailand, Oct. 2009 - Oct. 2009, pp. 111–115.
160. S. Dreiseitl and L. Ohno-Machado, "Logistic regression and artificial neural network classification models: a methodology review," *Journal of Biomedical Informatics,* vol. 35, 5–6, pp. 352–359, 2002, https://doi.org/10.1016/s1532-0464(03)00034-0.
161. R. Kumari and S. K. Srivastava, "Machine learning: A review on binary classification," *International Journal of Computer Applications,* vol. 160, no. 7, 2017.
162. A. Ghatak, *Deep learning with R.* Singapore: Springer, 2019.
163. V. Sessions and M. Valtorta, "The Effects of Data Quality on Machine Learning Algorithms," *ICIQ,* vol. 6, pp. 485–498, 2006.
164. M. I. Jordan and T. M. Mitchell, "Machine learning: Trends, perspectives, and prospects," *Science,* vol. 349, no. 6245, pp. 255–260, 2015, https://doi.org/10.1126/science.aaa8415.
165. S. Robinson, "Conceptual modelling for simulation Part I: definition and requirements," *Journal of the Operational Research Society,* vol. 59, no. 3, pp. 278–290, 2008, https://doi.org/10.1057/palgrave.jors.2602368.
166. J. L. Wahlers and J. F. Cox, "Competitive factors and performance measurement: Applying the theory of constraints to meet customer needs," *International Journal of Production Economics,* vol. 37, 2–3, pp. 229–240, 1994, https://doi.org/10.1016/0925-5273(94)90173-2.
167. W. J. Hopp and M. L. Spearman, *Factory Physics: Third Edition*: Waveland Press, 2011.
168. A. M. Law, "Simulation-models level of detail determines effectiveness," *Industrial engineering,* vol. 23, no. 10, pp. 16–18, 1991.
169. S. Robinson, "Sumulation projects: Building the right conceptual model project's objectives: The key to simulation modeling is to use the minimum amount of detail required to achieve the," *Industrial Engineering-Norcross,* vol. 26, no. 9, pp. 34–36, 1994.
170. R. K. Yin, *Case study research and applications: Design and methods*: Sage publications, 2017.
171. A. F. Agarap, "Deep Learning using Rectified Linear Units (ReLU)," Mar. 2018.
172. U. Fayyad, G. Piatetsky-Shapiro, and P. Smyth, "From data mining to knowledge discovery in databases," *AI magazine,* vol. 17, no. 3, p. 37, 1996, https://doi.org/10.1609/aimag.v17i3.1230.

173. R. Roscher, B. Bohn, M. F. Duarte, and J. Garcke, "Explainable Machine Learning for Scientific Insights and Discoveries," *IEEE Access*, vol. 8, pp. 42200–42216, 2020, https://doi.org/10.1109/ACCESS.2020.2976199.

174. B. W. Matthews, "Comparison of the predicted and observed secondary structure of T4 phage lysozyme," *Biochimica et Biophysica Acta (BBA) – Protein Structure*, vol. 405, no. 2, pp. 442–451, 1975, https://doi.org/10.1016/0005-2795(75)90109-9.

175. K. C. Green and J. S. Armstrong, "Simple versus complex forecasting: The evidence," *Journal of Business Research*, vol. 68, no. 8, pp. 1678–1685, 2015, https://doi.org/10.1016/j.jbusres.2015.03.026.

176. S. Chopra and M. S. Sodhi, "Supply-chain breakdown," *MIT Sloan management review*, vol. 46, no. 1, pp. 53–61, 2004.

177. J. Wagner, P. Burggräf, M. Dannapfel, and C. Fölling, "Assembly disruptions. Empirical evidence in the manufacturing industry of Germany, Austria and Switzerland," *International Refereed Journal of Engineering and Science*, vol. 6, no. 3, pp. 15–25, 2017.

178. F. Steinberg, P. Burggräf, J. Wagner, B. Heinbach, T. Saßmannshausen, and A. Brintrup, "A Novel Machine Learning Model for Predicting Late Supplier Deliveries of Low-Volume-High-Variety Products with Application in a German Machinery Industry," *Supply Chain Analytics*, p. 100003, 2023, https://doi.org/10.1016/j.sca.2023.100003.

179. G. Wang, A. Gunasekaran, E. W. Ngai, and T. Papadopoulos, "Big data analytics in logistics and supply chain management: Certain investigations for research and applications," *International Journal of Production Economics*, vol. 176, pp. 98–110, 2016, https://doi.org/10.1016/j.ijpe.2016.03.014.

180. T. Nguyen, L. Zhou, V. Spiegler, P. Ieromonachou, and Y. Lin, "Big data analytics in supply chain management: A state-of-the-art literature review," *Computers & Operations Research*, vol. 98, pp. 254–264, 2018, https://doi.org/10.1016/j.cor.2017.07.004.

181. A. Brintrup, "Artificial Intelligence in the Supply Chain," in *The Oxford handbook of supply chain management*, T. Y. Choi, J. J. Li, D. S. Rogers, T. Schoenherr, and S. M. Wagner, Eds., Oxford: Oxford University Press, 2020.

182. Y. Sheffi, *The resilient enterprise: Overcoming vulnerability for competitive advantage.* Cambridge, Mass., London: MIT Press, 2007.

183. K. Katsaliaki, P. Galetsi, and S. Kumar, "Supply chain disruptions and resilience: a major review and future research agenda," *Annals of operations research*, pp. 1–38, 2021, https://doi.org/10.1007/s10479-020-03912-1.

184. F. F. Rad *et al.*, "Industry 4.0 and supply chain performance: A systematic literature review of the benefits, challenges, and critical success factors of 11 core technologies," *Industrial Marketing Management*, vol. 105, pp. 268–293, 2022, https://doi.org/10.1016/j.indmarman.2022.06.009.

185. M. Christopher, *Logistics & supply chain management.* Harlow, England, New York: Pearson Education, 2016.

186. M. Christopher and H. Peck, "Building the resilient supply chain," *0957–4093*, 2004, https://doi.org/10.1108/09574090410700275.

187. G. Baryannis, S. Validi, S. Dani, and G. Antoniou, "Supply chain risk management and artificial intelligence: state of the art and future research directions," *International Journal of Production Research*, vol. 57, no. 7, pp. 2179–2202, 2019, https://doi.org/10.1080/00207543.2018.1530476.

188. M. A. Waller and S. E. Fawcett, "Data Science, Predictive Analytics, and Big Data: A Revolution That Will Transform Supply Chain Design and Management," *J Bus Logist*, vol. 34, no. 2, pp. 77–84, 2013, https://doi.org/10.1111/jbl.12010.

189. J. K. Bae and J. Kim, "Product development with data mining techniques: A case on design of digital camera," *Expert Systems with Applications*, vol. 38, no. 8, pp. 9274–9280, 2011, https://doi.org/10.1016/j.eswa.2011.01.030.

190. D. M.-H. Chiang, C.-P. Lin, and M.-C. Chen, "The adaptive approach for storage assignment by mining data of warehouse management system for distribution centres," *Enterprise Information Systems*, vol. 5, no. 2, pp. 219–234, 2011, https://doi.org/10.1080/17517575.2010.537784.

191. J. Cui, F. Liu, J. Hu, D. Janssens, G. Wets, and M. Cools, "Identifying mismatch between urban travel demand and transport network services using GPS data: A case study in the fast growing Chinese city of Harbin," *Neurocomputing*, vol. 181, no. 14, pp. 4–18, 2016, https://doi.org/10.1016/j.neucom.2015.08.100.

192. C. Ghedini Ralha and C. V. Sarmento Silva, "A multi-agent data mining system for cartel detection in Brazilian government procurement," *Expert Systems with Applications*, vol. 39, no. 14, pp. 11642–11656, 2012, https://doi.org/10.1016/j.eswa.2012.04.037.

193. C.-F. Chien, A. C. Diaz, and Y.-B. Lan, "A data mining approach for analyzing semiconductor MES and FDC data to enhance overall usage effectiveness (OUE)," *IJCIS*, vol. 7, Supplement 2, p. 52, 2014, https://doi.org/10.1080/18756891.2014.947114.

194. J. Wang and J. Zhang, "Big data analytics for forecasting cycle time in semiconductor wafer fabrication system," *International Journal of Production Research*, vol. 54, no. 23, pp. 7231–7244, 2016, https://doi.org/10.1080/00207543.2016.1174789.

195. C. Wang, X. Li, X. Zhou, A. Wang, and N. Nedjah, "Soft computing in big data intelligent transportation systems," *Applied Soft Computing*, vol. 38, no. 7209, pp. 1099–1108, 2016, https://doi.org/10.1016/j.asoc.2015.06.006.

196. P. Helo and Y. Hao, "Cloud manufacturing system for sheet metal processing," *Production Planning & Control*, vol. 28, 6–8, pp. 524–537, 2017, https://doi.org/10.1080/09537287.2017.1309714.

197. A. Kumar, R. Shankar, A. Choudhary, and L. S. Thakur, "A big data MapReduce framework for fault diagnosis in cloud-based manufacturing," *International Journal of Production Research*, vol. 54, no. 23, pp. 7060–7073, 2016, https://doi.org/10.1080/00207543.2016.1153166.

198. F. Bienhaus and A. Haddud, "Procurement 4.0: factors influencing the digitisation of procurement and supply chains," *BPMJ*, vol. 24, no. 4, pp. 965–984, 2018, https://doi.org/10.1108/BPMJ-06-2017-0139.

199. D. Ivanov, C. S. Tang, A. Dolgui, D. Battini, and A. Das, "Researchers' perspectives on Industry 4.0: multi-disciplinary analysis and opportunities for operations management," *International Journal of Production Research*, vol. 59, no. 7, pp. 2055–2078, 2021, https://doi.org/10.1080/00207543.2020.1798035.

200. M. M. Queiroz, D. Ivanov, A. Dolgui, and S. Fosso Wamba, "Impacts of epidemic outbreaks on supply chains: mapping a research agenda amid the COVID-19 pandemic

through a structured literature review," *Annals of operations research*, pp. 1–38, 2020, https://doi.org/10.1007/s10479-020-03685-7.

201. A. McAfee, E. Brynjolfsson, T. H. Davenport, D. J. Patil, and D. Barton, "Big data: the management revolution," *Harvard business review*, vol. 90, no. 10, pp. 60–68, 2012.

202. M. M. Queiroz, S. Fosso Wamba, M. C. Machado, and R. Telles, "Smart production systems drivers for business process management improvement," *BPMJ*, vol. 26, no. 5, pp. 1075–1092, 2020, https://doi.org/10.1108/BPMJ-03-2019-0134.

203. X. Zhu, A. Ninh, H. Zhao, and Z. Liu, "Demand Forecasting with Supply-Chain Information and Machine Learning: Evidence in the Pharmaceutical Industry," *Prod Oper Manag*, vol. 30, no. 9, pp. 3231–3252, 2021, https://doi.org/10.1111/poms.13426.

204. P. Liu, A. Hendalianpour, J. Razmi, and M. S. Sangari, "A solution algorithm for integrated production-inventory-routing of perishable goods with transshipment and uncertain demand," *Complex Intell. Syst.*, vol. 7, no. 3, pp. 1349–1365, 2021, https://doi.org/10.1007/s40747-020-00264-y.

205. S. K. Sardar, B. Sarkar, and B. Kim, "Integrating Machine Learning, Radio Frequency Identification, and Consignment Policy for Reducing Unreliability in Smart Supply Chain Management," *Processes*, vol. 9, no. 2, p. 247, 2021, https://doi.org/10.3390/pr9020247.

206. A. Hendalianpour, "Optimal lot-size and Price of Perishable Goods: A novel Game-Theoretic Model using Double Interval Grey Numbers," *Computers & Industrial Engineering*, vol. 149, no. 1, p. 106780, 2020, https://doi.org/10.1016/j.cie.2020.106780.

207. P. Liu, A. Hendalianpour, and M. Hamzehlou, "Pricing model of two-echelon supply chain for substitutable products based on double-interval grey-numbers," *IFS*, vol. 40, no. 5, pp. 8939–8961, 2021, https://doi.org/10.3233/JIFS-201206.

208. A. Hendalianpour, M. Hamzehlou, M. R. Feylizadeh, N. Xie, and M. H. Shakerizadeh, "Coordination and competition in two-echelon supply chain using grey revenue-sharing contracts," *GS*, vol. 11, no. 4, pp. 681–706, 2021, https://doi.org/10.1108/GS-04-2020-0056.

209. K. Nayal, R. D. Raut, M. M. Queiroz, V. S. Yadav, and B. E. Narkhede, "Are artificial intelligence and machine learning suitable to tackle the COVID-19 impacts? An agriculture supply chain perspective," *0957–4093*, ahead-of-print, ahead-of-print, p. 438, 2021, https://doi.org/10.1108/IJLM-01-2021-0002.

210. I. N. Pujawan and A. U. Bah, "Supply chains under COVID-19 disruptions: literature review and research agenda," *Supply Chain Forum: An International Journal*, vol. 23, no. 1, pp. 81–95, 2022, https://doi.org/10.1080/16258312.2021.1932568.

211. M. O. Alabi and O. Ngwenyama, "Food security and disruptions of the global food supply chains during COVID-19: building smarter food supply chains for post COVID-19 era," *BFJ*, vol. 125, no. 1, pp. 167–185, 2023, https://doi.org/10.1108/BFJ-03-2021-0333.

212. M. M. Bassiouni, R. K. Chakrabortty, O. K. Hussain, and H. F. Rahman, "Advanced deep learning approaches to predict supply chain risks under COVID-19 restrictions," *Expert Systems with Applications*, vol. 211, p. 118604, 2023, https://doi.org/10.1016/j.eswa.2022.118604.

213. P. Ghadimi, C. Wang, M. K. Lim, and C. Heavey, "Intelligent sustainable supplier selection using multi-agent technology: Theory and application for Industry 4.0 supply chains," *Computers & Industrial Engineering*, vol. 127, no. 5, pp. 588–600, 2019, https://doi.org/10.1016/j.cie.2018.10.050.

214. M. T. Ahmad, M. Firouz, and S. Mondal, "Robust supplier-selection and order-allocation in two-echelon supply networks: A parametric tolerance design approach," *Computers & Industrial Engineering*, vol. 171, no. 1, p. 108394, 2022, https://doi.org/10.1016/j.cie.2022.108394.

215. M. M. Khan, I. Bashar, G. M. Minhaj, A. I. Wasi, and N. U. I. Hossain, "Resilient and sustainable supplier selection: an integration of SCOR 4.0 and machine learning approach," *Sustainable and Resilient Infrastructure*, pp. 1–17, 2023, https://doi.org/10.1080/23789689.2023.2165782.

216. Du Ni, Z. Xiao, and M. K. Lim, "A systematic review of the research trends of machine learning in supply chain management," *Int. J. Mach. Learn. & Cyber.*, vol. 11, no. 7, pp. 1463–1482, 2020, https://doi.org/10.1007/s13042-019-01050-0.

217. K. Jensen, *Cross-industry standard process for data mining*. [Online]. Available: https://en.wikipedia.org/w/index.php?title=Cross-industry_standard_process_for_data_mining&oldid=1117200778 (accessed: Jan. 30 2023).

218. R. J. Hyndman and A. B. Koehler, "Another look at measures of forecast accuracy," *International Journal of Forecasting*, vol. 22, no. 4, pp. 679–688, 2006, https://doi.org/10.1016/j.ijforecast.2006.03.001.

219. G. G. Judge, R. C. Hill, W. E. Griffiths, H. Lütkepohl, and T.-C. Lee, *Introduction to the theory and practice of econometrics*, 2nd ed. New York: Wiley, 1988.

220. R. Bellman, *Dynamic programming*. Princeton, NJ: Princeton Univ. Pr, 1984.

221. H. Akoglu, "User's guide to correlation coefficients," *Turkish Journal of Emergency Medicine*, vol. 18, no. 3, pp. 91–93, 2018, https://doi.org/10.1016/j.tjem.2018.08.001.

222. C. Seger, "An investigation of categorical variable encoding techniques in machine learning: binary versus one-hot and feature hashing," School of electrical engineering and computer science, Royal Institure of Technology, Stockholm, Schweden, 2018.

Printed in the United States
by Baker & Taylor Publisher Services